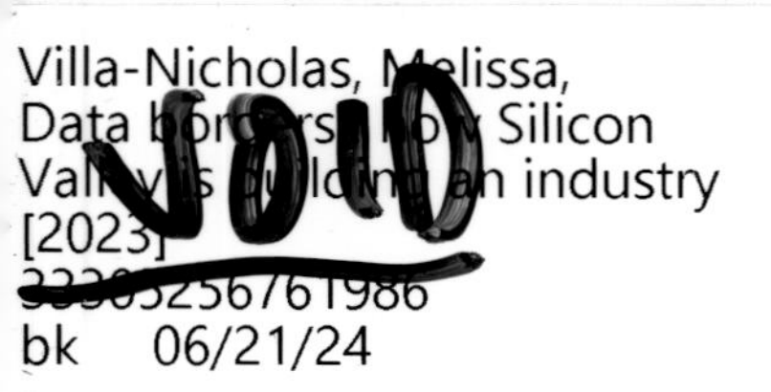

How Silicon Valley Is Building an Industry around Immigrants

Melissa Villa-Nicholas

UNIVERSITY OF CALIFORNIA PRESS

University of California Press
Oakland, California

Portions of this book were previously published in Melissa Villa-Nicholas (2020), Data body milieu: The Latinx immigrant at the center of technological development, *Feminist Media Studies* 20(2), 300–304, https://doi.org/10.1080/14680777.2020.1720351. Reprinted by permission of Taylor & Francis Ltd.

Library of Congress Cataloging-in-Publication Data

Names: Villa-Nicholas, Melissa, author.
Title: Data borders : how Silicon Valley is building an industry around immigrants / Melissa Villa-Nicholas.
Description: Oakland, California : University of California Press, [2023] | Includes bibliographical references and index.
Identifiers: LCCN 2022048753 (print) | LCCN 2022048754 (ebook) | ISBN 9780520386051 (cloth) | ISBN 9780520386075 (paperback) | ISBN 9780520386082 (ebook)
Subjects: LCSH: Latin Americans—Mexican-American Border Region—Social conditions—21st century. | Immigrants—Mexican-American Border Region—Social conditions—21st century. | Electronic surveillance—Mexican-American Border Region—21st century.
Classification: LCC E184.S75 V546 2023 (print) | LCC E184.S75 (ebook) | DDC 323.44/8209721—dc23/eng/20221107
LC record available at https://lccn.loc.gov/2022048753
LC ebook record available at https://lccn.loc.gov/2022048754

32 31 30 29 28 27 26 25 24 23
10 9 8 7 6 5 4 3 2 1

For Xochitl & Sofia, who show me the world through wonder

Contents

Illustrations

FIGURES

TABLE

BOX

PART ONE

The Data Body Milieu

Un Pincel de Rapunzel

This book starts with a parable. A reimagined techno-future. An undocumented Mexican woman's moment of freezing time and articulating something different. I don't want to start with the emerging collaboration of Silicon Valley and the Department of Homeland Security's vision of the US-Mexico border.

Instead, we begin here, at a table in Riverside County, California, eating *pan dulce* and *albondigas* during a mild January morning, when I interviewed family friends who came to our community decades ago, undocumented, and currently experience intense engagements with Customs and Border Protection (CBP), Immigration and Customs Enforcement (ICE), and surveillance in our borderland towns.

In asking people who are undocumented, beneficiaries of the Deferred Action for Childhood Arrivals (DACA) program, permanent residents, or naturalized citizens to speak their border-crossing stories, we also imagined techno-futures and alternate borders. What would they build, create, or imagine if it could be anything, without limits? Luz, a Mexican woman from Guanajuato in her fifties who had crossed the border over two decades previously, said: "*A mí me gustaría que existiría un pincel de Rapunzel.*"[1] She explained, "I would use Rapunzel's paintbrush so I could paint anything and go through the painting. I would go through to my mom's house [in Mexico]."

1. Names of interview collaborators have been changed to respect anonymity.

She had not seen her mom in over twenty years, since coming to the United States. As an undocumented person, she cannot travel to visit her family without extreme risk and danger. As you will read in chapter 6, her travel from Mexico was incredibly dangerous: she risked everything to give her children a better life. For a moment, she thought she had lost her children forever. The trade-off of crossing to the United States was leaving her mom and her younger siblings, whom she had also helped raise.

Luz imagines futures that allow her to walk through a painting. She imagines a portal that transcends space, time, citizenship, and borders to bring her back to her mom, without the restrictions of losing everything she has built in the United States.

This book anchors to her imagination and that of others like her, not the data borders I will be describing.

I can't interpret Luz's imagined futures more than I have already. To do so would take them away from what I call *first-person parables.* A parable is a story that, when heard, shifts the audience's course. The only way a parable can be one is if it changes readers' fixed perception so radically that they walk away different. First-person parables are about hearing life stories, dreams, and imagined futures from those who have been targeted for surveillance, detention, deportation, and marginalization by systems of control. First-person parables hail US immigrants as creators, innovators, and geniuses because they have traversed space and time to immigrate through bleak conditions, in contrast to majority narratives that hail technology CEOs as geniuses of our time. First-person parables are a counterdata initiative, as D'Ignazio and Klein call it in *Data Feminism* (D'Ignazio and Klein, 2020, p. 35).

First-person parables are how this book concludes. But this one is a placeholder in this introduction. A bookend for believing that methods of liberation are more resounding than marginalization and control.

Now for the introduction.

Introduction

FLAME OF THE WEST

In 2019, Palmer Luckey held a ribbon-cutting ceremony in Irvine, California, for his new technology defense company, Anduril Industries. Luckey had just come out of a tumultuous business relationship with Facebook, who had purchased his company Oculus, a tech company that came out with state-of-the-art 3-D virtual reality headsets. He was shifting his work toward a new vision: borderland defense technology. To cut the ribbon, Palmer had wanted to use his replica Lord of the Rings sword named Anduril that was carried by Aragorn in J. R. R. Tolkien's popular novels, but he didn't have time to sharpen it (Dean, 2019). Nevertheless, he drew on the Lord of the Rings mythos to convey the importance of this event:

> "Anduril," he said, leaning into the long Elvish vowels, "means Flame of the West. And I think that's what we're trying to be. We're trying to be a company that represents not just the best technology that Western democracy has to offer, but also the best ethics, the best of democracy, the best of values that we all hold dear." (Dean, 2019)

Anduril would go on to win a large bid from the Trump administration to bring together commercial technologies such as VR goggles, drones, and AI with the defense industry. Their experimental playing fields? The US-Mexico border. The aim? To capture immigrants crossing the border with the most advanced technologies developed in Silicon Valley, and to further build data profiles out of immigrant data.

This event symbolizes the evolving new threshold in borderland technology. The US-Mexico borderlands, always in cultural, political, and geographic flux, have shifted once again. Unlike the past, when new borderlands were drawn from the US-Mexico War of 1846–48, or the construction of a physical border wall, or burgeoning maquiladoras from the globalized economy of the mid-twentieth century onward, this change has cast the borderland as ubiquitous, digital, and often invisible to the eye. This emerging borderland stems from the partnership of various arms of the Department of Homeland Security, alt-right-leaning Silicon Valley startups, government agencies (such as state motor vehicle departments), and unfortunately for us, consumers like you and me.

I name this trend the *data body milieu*. Data body milieu is the state of borderland surveillance that brings all people, citizen and immigrant, into an intimate place of surveillance where our data lives together and defines us in a data borderland. It places Latinx immigrant data at the center of technological innovation and development.[1] In describing these data borders, I'm concerned with the liminal state in which almost every US resident lives: we cannot feel, describe, or point to when that data is in movement in favor of Immigration and Customs Enforcement (ICE), border patrol, and their parent the Department of Homeland Security (DHS). This interstitial border is always at play and yet rarely perceptible. In these emerging data borders, state, technological innovation, and data organization subjects coexist in a way that leads to the surveillance, capture, and deportation of undocumented people, without those subjects necessarily aware that they are interacting.

What does it mean when a liminal data border exists without space, time, or consciousness of those subjects' engagement? What does it mean when *everyone's* data interacts in order to police, apprehend, and deport across a border that is unseen?

A NEW VIRTUAL BORDER THRESHOLD

In March 2018, US Congress approved $400 million of the 1.3-billion-dollar budget for the 1,954 miles of virtual border wall, also known as a "smart" wall (Davis, 2019). It was estimated by the Office of Biometric Identity Management that DHS will be conducting 180 million biometric transactions a year among 260 million unique identities by fiscal

1. Word choices are explained in the "Terminology and Identities" section later in this chapter.

year 2022, with that number rising every year that passes (Homeland Advanced Recognition Technology, 2021). The virtual border wall was approved without the fiery debate about the physical border wall. But the rhetoric included promises that went beyond the physical border wall: Not only would immigrants be kept out of the United States, but they could now be known, documented through digital technology's biological mapping. The promise of the virtual border wall goes beyond the brick-and-mortar wall: It promises to solve *the Latinx immigrant threat* (Chavez, 2008)—a threat that reaches beyond the idea of citizenship in the United States into a source of anxiety concerning nonparticipation in producing data that is crucial for digital capitalism.

With the virtual border wall, technology accomplishes what ICE, the border patrol, white nationalists, English-only policies, Proposition 187, and voters in the borderlands could not accomplish over centuries of attempts to reverse the influx of the Latinx population in the US borderlands; a promise of technological futurity that arose with more gusto in the 2010s, when border technology proposed a United States with a controllable immigrant influx at the border.

We are increasingly seeing Latinx immigrants in borderlands referred to as data and engaged as the object of mobilizing information technology and defining citizenship inclusion. Recent investment in the collection of biodata on the border and around the "belonging" of citizenship is a highly profitable grab around different groups of immigrants, Latinx undocumented people, permanent residents, and Latinx citizens (Cagle, 2017). The surveillance of Latinx immigrants and development of technology around Latinx bodies is not new (Chaar-López, 2019); but the scale and networked circulation of that data has changed. As data gathering increases, US citizens and Latinx immigrants become more intertwined in the borderland milieu that historian Oscar Martínez originally theorized, into what is now a state of data body milieu.

This book is about the emerging state of borderland surveillance that brings all people, citizen and immigrant, into an intimate place of surveillance where our data lives together and defines us in a digital borderlands. This surveillance places the Latinx immigrant body at the center of technological innovation and development and an emerging industry at the crossroads of Silicon Valley and ICE. Companies such as Quanergy Electric, Anduril, BI2, Palantir, Amazon, LexisNexis, and DNA testing companies all have a stake in gathering data of undocumented people at ports of entry, borderlands, detention centers, and immigrant-populated cities—and subsequently US citizens as well. While surveillance and contentious relations along the US-Mexico border are not new, what is new is both the scale at which data is gathered and the move to biological data—from retina scanning to DNA testing.

This is the evolving state of data body milieu. Latinx immigrants becomes valued as a data body, one that is used for purposes of technological design and valuable as a source of data in and of itself. Silicon Valley is physically reshaping around the US-Mexico borderlands, and US residents engage in a constant state of borderland surveillance, intimately entangled with undocumented data surveillance. Information technology on the border and in the ubiquitous data borderland is approached as an avenue to manage the excess of Latinx immigration into the United States. This emerging industry posits that the "next new" technology can contain the US-Mexico border's rugged terrain, quantify immigrants under control, and manage the nonassimilated excess of Latinidad.

Silicon Valley's move to design technology around Latinx immigrants is building on a long history of surveillance projects networked into and justified around communities of color as a perceived threat to white and citizen safety. Data body milieu is the name I give this recent trend, but it is always interconnected and built onto the ways in which surveillance and technologies have been encoded with bias, racism, sexism, classism, and ableism to benefit normative and acceptable states of citizenship.

This book does not encapsulate all immigrant experiences across the United States and is not comprehensive of the Latinx immigrant experience. I'm focused on the US-Mexico border, the Latinx immigrants traditionally targeted in political rhetoric by way of US anxieties along that border (Mexican and Central American), and the communities built around migrations and residencies from that launch point. There should be work that focuses further on the ways in which surveillance technologies and the developing data body milieu sets its gaze on different immigrants, and how these forms of policing are interconnected.

Black, Latinx, Middle Eastern, Indigenous, Asian American, and Muslim communities have been used as a launching point and a central focus for developing surveillance technologies for decades. One of the most infamous of those was the COunter INTELigence PROgram (COINTELPRO), the US FBI project that surveilled American political organizations from the 1950s through the 1970s, with a particular focus on Black, Chicano, and Native American organizations (Ogbar, 2017). Post 9/11, the 2001 Patriot Act allowed US law enforcement to use extreme surveillance techniques to investigate terrorism-related crime; it allowed for wiretapping, the collection of American phone records, delayed notification of search warrants, federal agents to obtain bank and business records, network information sharing between government agencies, and further screening in travel and airports (Puar, 2007). That act brought surveillance into the public sphere as far as the public library, with librarians receiving National Security Letters: gag orders by the FBI to turn over patrons' book checkout records (Chase et al., 2016).

Another group drastically impacted by these developed and emerging surveillance technologies are the Indigenous nations that live along the border. Native Americans live and have lived in the ever-changing borderlands since before colonization, and they experience surveillance projects intensely themselves. Most recently, the Integrated Fixed Towers, built by Elbiet systems, Israel's military defense company, have been forced onto the Tohono O'odham Nation in Arizona. Those Indigenous groups are harassed by CBP daily, and their own lives have changed in the span of one lifetime, the borderlands so militarized that their previous more fluid movement and community from Mexico to the United States has lost its flexibility and severed their previous community network (Jaacks, 2020; Parrish, 2019).

Many scholars have described the moving parts of the current power structures of technology and society. In 2016 Cathy O'Neil raised the red flag from her own work in data processing and mathematics, finding that these mathematical algorithmic systems were determining high stakes in citizens' everyday lives, such as which teachers to fire based on numbers and algorithmic decisions, not their abilities; health care access and cost based on an individual's likelihood to get sick; and stop-and-frisk policies, to name a few.

Internet scholar Safiya Noble (2016) observed the seemingly "neutral" technologies of our everyday lives as deeply intertwined in the power dynamics that are embedded in our social structures. Noble lifted this veil by naming these "algorithms of oppression": she searched for "Black girls" on Google and found solely pornographic images on the first page of results of the web and image search; she found the same for Latina and Asian girls (Dave, 2022, Noble, 2018). She described the bigger picture here: technologies in our everyday lives are not neutral and value free but indeed reflect the anti-Black and misogynist social structures that have been established in the United States. Now that same racism, sexism, and class inequality is built into algorithms—algorithms that determine home loans, medical coverage, and other everyday life necessities. African American Studies scholar Ruha Benjamin (2019) calls the benign and often altruistic ways in which these technologies are developed, designed, and delivered into consumers' hands the *New Jim Code,* justified through the necessity of progress. We will see those justifications used to embed data border technology into everyday systems of information frequently throughout this book.

On the topic of data borders, I am in conversation with scholars in various fields. These include critical internet and media scholars who interrogate how information technologies reproduce and challenge power and resist values holding that technology is neutral or technology leads to democratized progress. Latinx studies scholars have looked at the ways in which the state grants and denies citizenship based on race, gender, and sexuality. And library and information science (LIS) scholars think about ways in which information and data is organized that reaffirms structural inequality, presenting responses to further information equity instead. I also benefit from and discuss work on the quantified self, research on new media studies, the digital humanities, and histories of computing and technology.

The contemporary surveillance state is a messy network, like that box of old electronic cords that you have in your garage. This immigrant data surveillance state is a large ball of tangled mess that works together to connect and network in data that determines everything from our medical coverage and eligibility for loans to our movement across borders. This book attempts to pull on some of those wires and untangle this mess that is the contemporary surveillance state that organizes around Latinx immigrants along the US-Mexico border. I circle around questions such as: How are most people in the United States now connected to ICE systems of surveillance? How are technologies designed around Latinx immigrant data? How are US residents' data bodies living outside of our physical bodies? Also important to this study is: How are most people experiencing a borderland by way of their data, consciously or not?

One purpose of this book is to promote and accelerate immigrant data rights as a part of new necessary movements for immigrant rights overall, by demonstrating what this intimate digital surveillance state centered on Latinx immigrant (and perceived immigrant) data looks like, how it operates, how it builds on what came before and moves beyond, how it classifies and categorizes, how it expands beyond just

Latinx people, how it is commercialized and consumed. I weave personal stories of growing up in the physical borderlands as a second-generation Mexican American Latina to illuminate the contrast of the disembodied and embodied data borderlands. I'm concerned with the question: What does it mean that many people's data is in a constant state of correlation to ICE systems of surveillance, but they can't feel those borderlands? I continually reflect on my experience growing up in borderlands, and I bring in Latinx immigrant experiences of the borderlands to both put the body back in the data body and contrast the data border experience that is so pervasive in everyday lives.

My intention is to describe what is going on with the emerging commodified surveillance state and push toward immigrant data rights. But my hope is to tell the story of how I ended up networked into immigrant surveillance. The story of how you are networked into immigrant surveillance and deportation. The call to action for this story is to lead with undocumented immigrant data rights in policy by pivoting with parable in the concluding chapter, by ending with immigrant experiences of the borders and imagining techno-futures. Pivoting from story, I hope immigrant experiences of crossing the border, their awareness of the constraining technology of surveillance, and their imaginings of alternate borderland techno-futures act as a parable to counter the larger story structure in which we find ourselves (described further in chapter 5).

A second aim of this book is to look at what is happening around Latinx immigrant inclusion in and exclusion from the state by way of emerging information technology mergers between Silicon Valley's collaboration with the Department of Homeland Security (particularly ICE and border patrol). Not all forms of Latinidad are rejected by the state. For example, the now defunct Latina-designed AI "Emma," a chatbot that once faced immigrants on the United States Citizenship and Immigration Services (USCIS) website to answer questions about the citizenship process. Emma, and other emerging Latina AI, demonstrates that there are acceptable forms of Latinx immigrant assimilation and citizenship, when that citizenship is done "right," with features that are more Anglo, speaking English first and Spanish second (Sweeney and Villa-Nicholas, 2022; Villa-Nicholas and Sweeney, 2019). The emerging "datafication" (quantification of everything) trend toward Latinx immigrants in the borderlands suggests that those not immigrating through legal means can be removed back to their home countries and their citizenship "set right" with quantifying tools.

METHOD

This book uses public-facing information, tech companies' contracts with ICE, government documents, mainstream media coverage of the evolving data borders, leaked government files, advertising, and other electronic resources as observational fieldwork, along with interviews from Latinx people of varying citizenship status from my hometown.

I draw on multiple interviews with a small group of people from my hometown and surrounding areas. I interviewed fourteen Mexican people over the past five years who are from Mexican immigrant communities, including undocumented, DACA Dreamers, permanent residents, naturalized citizens, and those born into citizenship. I use these interviews in chapter 6 on reimagining borderlands and techno-futures. Valuing qualitative methods as "small data" embeds agency in the interviewees as a means of *counter-storytelling* in the face of the larger stories positioned by Silicon Valley-ICE collaborations. LIS scholar LaTesha Velez writes, "Counter-storytelling [is] a means for minorities to share their experiences and to counter existing hegemonic narratives (Delgado, 1989; Solórzano and Yosso, 2002)" (quoted in Velez, 2022). Counternarrative, counter-storytelling, and counterdata, conveyed through my own experiences and those of Latinx immigrants, work to interrupt the dominant narrative that is built into Western values around technology as progress and Silicon Valley story structures of the heroes' journey, a topic discussed more in chapter 5 (D'Ignazio & Klein, 2020; Velez, 2022). This work is also woven in with my personal autoethnography of growing up Mexican American and geographically close to the San Ysidro port of entry, to contrast and explain the corresponding data borders. Autoethnography too works as a form of counternarrative to dominant Western values (Velez, 2022, p. 209). I engage these methods intentionally to value small data and resist the overreliance and overfunding of big data that has entrenched academic and industry data collection methods.

TERMINOLOGY AND IDENTITIES

Terminology is important and often slips at the intersections of the themes throughout the book. I use the word *Latinx* to describe people of Mexico, Central and Latin America, using the *x* to acknowledge that not all genders fall into the *-o* or *-a* category of male or female binaries. Many Latinxs do not identify with this term and do not come to a consensus with one term in general. Latinx immigrants differ in race, gender, sexuality, languages, religion, nationality, class, and especially in how they/we identify with naming terminology. The emerging popular term *Latine* is also a favorable nongendered term that is more smoothly incorporated into Spanish speech.

I use *data borders* as a term to encompass the many layers of information technologies that are building the physical borders at the US-Mexico border and that extend around the United States through data and digital surveillance. Throughout the book, I will describe various types of information technologies that include physical infrastructure applied on the US border; artificial intelligence (AI) that may be built in a lab in California's Orange County but applied along Texas, Arizona, and California; and the harvesting of data from many governments and private sector sources. Ultimately, I favor the term *data* to modify border over *information technology*, *digital*, or *smart* because data has become the underlying valued source of capital in this phenomenon. This isn't to say that physical infrastructure isn't always already present, as I will describe in chapter 1, but that the "networked society" that sociologist Manuel Castells (1997) names has commodified information as data as the primary source of capital, especially with regards to the Latinx immigrant population.

A MAP OF THIS BOOK

Part I describes and analyzes the emerging state of data borders and data body milieu. Chapter 1 discusses how the US-Mexico border became a borderland as a geographic place. The US-Mexico border, built through wars, mutual agreements between the two countries, a series of immigration policies, and anti-immigrant sentiments in the United States, is still in flux today. After establishing this historical premise, I look at how the US-Mexico border expands throughout the country with new surveillance technologies. Silicon Valley companies establish new data frontiers through their own relocation into borderlands. The US-Mexico border is approached as the new ground zero lab for technological ingenuity and futurity.

Chapter 2 engages with how the Latinx immigrant has become the center of technological design in these new virtual borderlands. Surveillance technologies promise to tamper the Latinx threat by way of data collection, storage, and retrieval. For Latinx immigrants who do not meet the standards of the ideal US citizen, full documentation of their data body is required. I look further into the new trend of biologically mapping undocumented people through biotechnologies or biometric scanning technologies. There is an increasing and feverish trend to document the whole body with these biometric technologies. As tech consumers volunteer their genetic data, face, retina, and fingerprint scans for full-privilege benefits, we see the inverse happening with immigrant data—a competition among technology companies to biologically scan Latinx data involuntarily. Through this chapter, I argue that Latinx immigrants are increasingly valued for their data as a commercial product.

Chapter 3 unweaves the tangled web of immigrant information surveillance networks. Those contemporary technologies, proposed for innovation, are built upon systems perfected around the surveillance and control of communities of color. This data body milieu is not a stand-alone phenomenon but a network, dependent upon older physical infrastructures and databases that gather data from marginalized communities and make suggestions on how to act on that data, for purposes of exploitation and control. This network includes the surveillance of communities of color, incarceration, law enforcement, border patrol, ICE, and citizen consumers, to name a few. I describe the network of commercial, military, public, and private information systems as the new *Migra* (border patrol).

Chapter 4 examines how the prototypical "good citizen" engages in these surveillance and data collecting strategies around Latinx immigrants. Private and public data and physical spaces have become networked into ICE databases, as this US American becomes a part of correlated ICE surveillance and deportations through their personal data and public participation. In this chapter I will dive deeper into how this data sharing impacts the public, such as public service employees and their patrons, by way of librarians and libraries, who tread their own use of data with uncertainty as they have found that some of their database subscriptions, such as Elsevier and LexisNexis, are networked into ICE data mining. The public library and librarians are in many ways a demonstration of how public entities find themselves engaged in this data dragnet, as well as the uncertainty and gray area that comes with using learning technologies. Networked belonging through correlated data *makes* citizenship and contributes to building immigrant data subjects.

Chapter 5 looks at how Silicon Valley uses storytelling to make their companies relatable to US citizens, building new borderlands through their use of science fiction, fantasy, and video game references. Technology companies draw on the power of story and the heroes' journey to pit the "good guys"— tech startups, giants, designers, coders, and users—against the "bad guys"—undocumented immigrants as well as Black and brown communities around the country networked with those immigrants. Storytelling and mythologies are the invisible Western morals that hold these data borders up.

Part II is the beginning of a conversation around responses, rejections, and reimagining this data border state. Chapter 6 imagines different borders and information technologies with people from my community in Riverside County that are undocumented, DACA recipient, permanent resident, naturalized, or first-generation Latinxs, all of whom have interacted with surveillance at the California borders and border patrol. While interviews with these collaborators are mixed throughout the book to speak back to the ubiquity and invisibility of surveillance technologies, this chapter in particular uses imagination as a method of response to the data borderland. I approach imagining different systems as a counterweight to chapter 5's powerful mythos of storytelling. Rather than a story, I hope to promote parables of reimagining techno-futures and alternate borderlands.

This book draws heavily on mainstream media, government documents, leaked documents, Silicon Valley companies' media, and qualitative interviews. It looks at the construction of Latinx immigrants as sources of data, gathered and constructed by media, tech companies, US government entities such as ICE and CBP, and other commercial and educational organizations. I interview Latinx people of various immigrant statuses from my hometown and surrounding areas who reside in a highly surveilled borderland and have had varying degrees of interaction with border patrol, ICE, and detention centers. Chapter 6 especially aims to build agency into the discouraging scale of the surveillance state. The book as a whole proposes to name this emerging state "data body milieu." Therefore, often I prioritize calling out the ICE-Silicon Valley hybrid's "story," but chapter 6 attempts to balance those scales.

I conclude by exploring methods of resistance to this state of data body milieu. My aim is to provide agency on a bleak topic that has seemingly become endemic and unraveled in front of our eyes in the twenty-first century, by advocating for immigrant data rights. In the conclusion I consider the multitude of movements working toward immigrant data rights: Latinx students at Dartmouth College, librarians across the country, community organizers, and policy special interest groups. For example, Latinx organization Mijente's *Tech Wars* webinar and immigrant data rights activism organizing series, which is designed for "anyone interested in studying technology and data as the new frontier in organizing against the systems of enforcement and criminalization that harm our communities" (Mijente, 2022). Academics, organizations, and budding policy are a rising tide in immigrant data rights activism. Where there are walls, there are tunnels.

To understand this new state of data borderlands, let me first describe the borderland milieu.

A CHANGING BORDERLAND MILIEU

In 2014 a busload of immigrant adults and children, already detained with border patrol, were en route to Murrieta, California. They were to be housed in the Murrieta detention center and then united with their families throughout the United States, monitored by ankle bracelets.

They were met by a group of protesters with signs that read "Return to Sender," and "Stop Illegal Immigration." Protesters and counterprotesters lined up facing one other, and some demonstrators physically stood in front of the bus to block its entrance into the detention facility. To many Latinxs' dismay, singer and local resident Lupillo Rivera (brother to the deceased Mexican singer Jenni Rivera) was spit on by a protester. Latinx counterprotesters had their own signs of *bienvenidos* (welcome). Those buses were sent to a facility in San Diego instead.

I was born in Torrance, California, and my family moved to Murrieta when I was nine. I'm second-generation Mexican American and mixed—my mom is Mexican American, my dad is Anglo American. The tension of racism in Murrieta toward Mexicans was palpable throughout my life. Growing up in a borderland, less than one hundred miles from the border, it was not uncommon to see those white vans and SUVs with the green stripe across them: border patrol, migra. It was not uncommon to see police or border patrol pulling over Mexican men on their way to work in their dungarees, long-sleeved work shirts, and caps. Their truck beds framed by planks of wood, with shovels, spears, weed whackers, and other tools all standing straight up, indicate that those men are landscaping around the spiraling, pink-and-beige tract homes of the suburb.

If you're driving from San Diego to Murrieta on Interstate 15 northbound, you would stop at the Rainbow border checkpoint just before Temecula. The hills rising above the checkpoint are dotted with Southern California shrubs. Sage lingers in the air in the Inland Empire. Dew dots across the fresh-cut grasses of many lawns every morning, and overcast settles from the Pacific Ocean just on the other side of the mountains that lie between Riverside County and San Diego.

If you grew up in Riverside County, then the signs, smells, and landscapes of the border were familiar, and surveillance was familiar. You know what it felt like.

Also familiar are trucks piled high with furniture, clothes, and toys, heading to Mexico after a good yard sale to resell products, or making the long drive down to see family deep in the heart of Mexico. It's common to load the truck or van up with everything from clothes to shoes to toys for family in Mexico.

In the 1990s, I remember Prop 187 passing, the proposition that prohibited undocumented immigrants from using health care, education, and other services in California. I was in middle school and there were suddenly kids missing at school. At tables and spread across the open, cement-based quad, kids were just gone.

Once, when I was a teenager, my mom was stopped and questioned by border patrol. I remember the dry heat making my legs stick to the fraying leather seats. The foam from the inside of the seats seeping from the seams. I was scared, though my mom wasn't undocumented. Still, I didn't know the extent of the border patrol agent's power. "US citizen," she repeated over and over to his questions. I can't remember the questions. Just the answer on repeat, "US citizen." The smell of sage on a desert night floated through the open window. The Southern California mountains visible in outline.

When I was growing up in this valley, deportations and raids were common. Still, the Mexican, Salvadorian, and Guatemalan populations continued to grow, and the valley settled down into a cultural blend, always intimate, always tense. Non-Latinxs eat traditional Mexican food beyond Taco Bell. They seek out albondigas and tostilocos. Piñatas went from rare to a norm at birthday parties.

Cultures mixed, resisted, and strained next to and among one another.

In 2010 in Arizona, Senate Bill 1070 was proposed to justify racially profiling all Latinxs to seek out and deport undocumented people. Throughout 2010, Temecula, Murrieta, and Lake Elsinore passed various ordinances supporting legislation similar to SB 1070, such as E-Verify requirements. E-Verify is an employer checking program that checks into citizenship status and has long been supported by the anti-immigrant movement. A deeper look into the anti-immigrant activists in these borderlands who pushed this legislation forward in the Temecula Valley demonstrates long ties to alt-right groups (ADL, 2014).

Once, in 2017, I went to Cardenas, our major Mexican supermarket, to buy some small Tajín bottles for stocking stuffers. An Anglo woman behind me in line said, "Oh those are great, I get them for my family too." My boyfriend laughed and said, "I guess they're coming around to us."

But this is not the data body milieu, the emerging state that this book names and attempts to describe.

What I'm describing is a borderland milieu, a term that Oscar J. Martínez (1994) coined in his deep study of how people live, clash, and interact along the US-Mexico border. They form relationships, culture merges, sometimes culture is appropriated. They have tense and fraught frictions with each other. They adopt each other's values, they have conflicting, sometimes violent resistance to each other's cultures. Borderland milieu, as defined by Martínez, is "the unique forces, processes, and characteristics that set borderlands apart from interior zones including transnational interaction, international conflict and accommodation, ethnic conflict and accommodation, and separateness. In their totality, these elements constitute what might be called the borderlands milieu" (1994, p. 10).

Borders are geographic locations, but geopolitics also imagines them and makes them tangible. The US-Mexico border had not been the focus of immigration flow until the 1920s (Ngai, 2004, p. 64). Immigrants are also valued or rejected based on politics, race, gender, class, and sexuality. When the US-Mexico border became more of a tangible line, patrolled and regulated, it was monitored for illegal entries of Chinese immigrants by way of Mexico. Borders and the people that are excluded or included are often constructed, physically and in the social imagination, by way of immigration policy. The insiders and outsiders of these "imagined communities," to use political scientist Benedict Anderson's (1991) term, make up nations and nationalism.

Scholars argue that borders are not always a geographic location but rather include consciousness and identity formation; as Gloria Anzaldúa (1987) has observed, Latinx immigrant communities' borderland experiences have multiple meanings, including deeply personal and intimate experiences of living in an in-between state. Borderlands engage with the material and social exchanges and connect to individual subjects' experiences of intersectional inequities, personal belief systems, and living in interstitial cultures. According to Segura and Zavella (2007), it's also "feeling 'in between' cultures, languages, or places. And borderlands are spaces where the marginalized voice their identities and resistance. All of these social, political, spiritual, and emotional transitions transcend geopolitical space" (p. 4). Latinx immigrants have long navigated borderlands in their everyday lives, through migration but also through interpersonal cultures and intimate spaces. This *in-between-ness* (Paredes, 1993) emerges often in Chicanx/Latinx thought around identity, migration, and border theory, or what Chicana studies theorist Angie Chabram-Dernersesian calls the multi-focality of Latinx immigrant communities' experience in the United States. Anzaldúa (1987) named borderland consciousness as a constantly shifting process that allows for contradictions and new identities from hybrid cultures.

It's difficult to see this milieu when you grow up in it; it's the quotidian state of things, while also place based and identity forming. I suppose that comfort in this state also made it difficult to see that something had changed over the past few years.

"When did they build a border facility in Murrieta?" I ask family and friends, unable to remember the detention center nestled in next to a church and preschool. The borderland state naturalized itself once again.

Just as the data body does.

DATAFIED

The data body is another state that we've settled into quite comfortably. There was a time when we weren't sure if we should give our phone number to the grocery store for a reward system. Or whether to provide our emails to sign up for accounts. Or hesitated to sign up for online billing.

Everything went online. And occasionally we said, "I just hope this doesn't get into the wrong hands." Meanwhile many hands in tech's coding and design sectors built systems that exacerbated the current states of inequality and correlated databases. And that data led to a massive network with many entrances that could surveil and deport immigrants.

By now, the data body has not only normalized, it's become routine.

Felix Stalder (2002) first referred to the data body as a "shadow body" that follows us through classification and measurement. It increased with every digitized byte of data from citizen consumers' credit cards, grocery purchases, and online bill payments (Stalder, 2002). Of course, this data body comes with a series of rewards and necessities that give us chances to improve our standard of living, as with health insurance and mortgages. Rewards programs, of course, manifest just by giving out an email address or phone number in shopping. But this data body also networks into the same systemic oppressions of race, gender, sexuality, and class that Safiya Noble (2018) has named "technological redlining," a result of algorithm bias, making it in fact deeply bodied. She finds that "There is a missing social and human context in some types of algorithmically driven decision making, and this matters for everyone engaging with these types of technologies in everyday life. It is of particular concern for marginalized groups . . ." (p. 10). The data body, then, echoes and often exacerbates the inequity and stratifications of the physical body. The data body precedes itself, "Our bodies are being shadowed by an increasingly comprehensive 'data body.' However, this shadow body does more than follow us. It does also precede us" (Stalder, 2002, p. 4). Technological redlining affects the access to quality-of-life improvements such as health insurance, loans, and credit cards, as well as the greenlight to easily travel across borders (Noble, 2018; Puar, 2007, p. 155). As the activist and community organization the *Our Data Bodies Project* asserts, "our data body can also follow behind us like a digital shadow, impacting the choices and opportunities we face in the future. Though our data bodies are mostly digital, the collection, analysis and sharing of data have a profound impact on our ability to meet our basic material needs: food, shelter, medical care, safety, and employment" (Saba et al., 2017, p. 3).

This data body is arranged by what Noble calls algorithms of oppression, which she has theorized around the Google algorithm, and how those manifest and recirculate racism. That form of the data body mirrors and replicates the larger systems of racism, sexism, gender discrimination, and class divides. The justification of building more surveilling technologies happens through Benjamin's New Jim Code, wherein technologies that harm Black communities are justified through their efficiency and ability to move society forward. About the New Jim Code, she says it involves "the employment of new technologies that reflect and reproduce existing inequities but that are promoted and perceived as more objective or progressive than the discriminatory systems of a previous era" (Benjamin, 2019, p. 6). For example, those who claim facial recognition software will lead to catching the right criminal or terrorist in cities and airports all over the country neglect the fact that those technologies focus on Black and brown people and that registering their data leads to further profiling databases. Algorithms that are sold as altruistic and neutral, or even "fixable" when they reproduce racial bias, become instead the standard of surveilling, classifying, cataloging, and retrieving the data body, all organized by race, gender, sexuality, class, and other intersections. These technological assemblages are co-constructing data borders; legal scholar Ana Muñiz builds on criminologists' work and argues that bordering is a process, and that digital forms of surveillance in borderlands are a part of this process (2022, p. 4).

Part of the process of how data is made, and then gathered, is how it is organized. What is the difference between information and data? Data is everywhere and everything. It is created in many forms, including verbal, visual, and written—it is physical and digital (Bawden & Robinson, 2013, p. 53; Olwagen, 2015). Moreover, it is flowing at incredible speeds through internet access and connections. When organized and given meaning intentionally, data becomes information. LIS professor and one-time cataloging librarian Hope Olson wrote about *the power to name* in information collection practices. Power and authority are in action when data is collected and assigned names and meaning, when that information is *named* and classified into information. Anyone who gathers information and organizes data implements bias (Olson, 2002).

Data too has its own history, context, and assumed role as neutral, though it is anything but. Data has always been *made*. Data is not a natural resource, and like technology, it is not neutral. The infrastructure for making data is also not male. Many researchers have written on how there was a time when women *were* the computers. Data and quantification, as D'Ignazio and Klein argue, have a historical relationship with power. This is datafication, the process of quantifying literally everything. D'Ignazio and Klein (2020) demonstrate how governments and corporations utilize data for methods of control and normativity (p. 17). Data is *made* by way of many intersectional identities that are shaped by and shape power—race, gender, sexuality, class, ableism, citizenship—leading to what Noble (2018) demonstrated on how Google searches codify race and gender through searches and algorithms. Systemic inequities around intersectional identities are built into technological systems, and built into making data, by way of technology design. Data can be and is often made from and within the bias and inequity of the United States and throughout the world, but it can also have counternarratives, which I employ in chapter 6 and the conclusion. I attempt to use counternarratives throughout the book to demonstrate that Latinx immigrants and citizens in the United States have technology histories and critical technological practices.

Privacy and surveillance have been concerns among Americans, paralleling anxieties around citizenship and data collection of immigrants. Sara Igo writes that privacy is difficult to define. It is political and cultural in that US Americans have contended with how much they want to be *known* citizens (Igo, 2018, p. 2). While contemporary ongoing debate about privacy may be a given, modern privacy has not always been a public importance to Americans. In the nineteenth century, people began to speak about the right to privacy, and in the twentieth century privacy rights took hold (Igo, 2018). "Privacy in the modern United States has been less a thing with definite contents than a seedbed for social thought, a tool for navigating an increasingly knowing society" (p. 16). Igo outlines various times in US history that American citizens were thinking about and identifying "privacy" and "being known." She states that the first modern call for a right to privacy occurred in the 1890s Victorian era. With the turn of the twentieth century, the public's anxiety heightened around the growing government's practice of cataloging citizens. In the 1960s, as technology developed, privacy became a constitutional right. In the 1970s and after, there were political demands for "transparency, exposure, and disclosure" (p. 13), and into the twenty-first century, larger perceptions of the commercialization of surveillance concluded that there "was no longer any privacy in the United States, nor even any desire for it" (p. 13).

There are many layers and entryways into citizenship. Igo addresses how privacy has been unequal since the 1900s depending on citizenship status: "As a general rule, those excluded from full political citizenship because of their class, race, gender, age, nationality, able-bodiedness, or sexuality—or combination thereof—also suffered most from a lack of privacy" (Igo, 2018, p. 9). Citizenship and privacy have long gone hand and hand. Debates over privacy happen often as social anxieties are shifting about a larger cultural or national change (Green, 2001; Igo, 2018; Villa Nicholas, 2018). A history of privacy from within and around the periphery of citizenship and immigrants of varied status is intertwined with American histories of privacy:

> Across the last century privacy was increasingly linked to that most public of identities, the rights-bearing citizen. Privacy talks thus became a potent avenue for claiming *and* circumscribing the social benefits of a modern industrial democracy. At the same time, instating on recognition—as a citizen, a holder of a specific identity, a person out of the shadows—was basic to

> enacting one's membership in society. How well known a citizen would be was a sensitive marker of status and power, a fissure like another cutting across professions of equality and opportunity in American life. (Igo, 2018, p. 10)

Immigrants, especially Latinx immigrants, are always, as Igo says, the shadow in American privacy debates, anxieties, and negotiations.

Latinx immigrant communities enter this privacy debate and data body state regarded as outsiders to information technology benefits and skills, while simultaneously deeply surveilled and highly visible products of state formation by way of surveillance. Over the years they have straddled the tensions of information technologies by being assumed to be nontechnological and by way of social inequities, often excluded from some of the more sophisticated information technologies and their corresponding jobs in science, technology, engineering, and math (STEM) fields. Mexican artist and activist Guillermo Gómez-Peña noticed this trend with the increasing public use of the internet in the 1990s: "The mythology goes like this. Mexicans (and other Latinos) cannot handle high technology. Caught between a preindustrial past and an imposed postmodernity, we continue to be manual beings—homo fabers par excellence, imaginative artisans (not technicians)—and our understanding of the world is strictly political, poetical, or metaphysical at best, but certainly not scientific" (2001, p. 192).

However, Latinxs have been the backbone in the physical labor of building information technologies, necessary to the infrastructure; from borderland maquiladoras that build much of the world's electronics to Latina telephone operators (Funari & De La Torre, 2006; Peña, 1997; Villa-Nicholas, 2022). Their labor is highly invisible, assisting in shaping the state and strengthening technological Western values (Villa-Nicholas, 2022). Latinx immigrants and US residents enter this data body as both producers of the information technologies that quantify subjects and products of these systems; both included in biopolitical state formations by labor, consumption, and good citizenship *and* surveilled and excluded by developing sophisticating systems of data gathering.

Along the US-Mexico border there is a tradition of building surveilling information technologies around what Latinx studies theorist Leo Chavez has named the "Latino threat" (Chavez, 2008). Through his history of the electronic fence along the border, borderland technology scholar Iván Chaar-López (2019) demonstrates that "actors and machines traced the boundaries of the nation on the ground and on human bodies imagined as intruders," finding that people crossing the border have been historically regarded as data. Chaar-López asserts that state building happens by way of technology drawing a racial line of exclusion and inclusion around "undesired" Latinx immigrants. In 2019, *Intercept* writer Max Rivlin-Nadler wrote a story that detailed how networked information from social media, the Department of Motor Vehicles (DMV), daycare facilities, hospitals, and more was leading to the deportation of undocumented people, referring to it as "the ever-widening surveillance dragnet ICE uses to track down immigrants who are subject to deportation." In this case, ICE used Thomson Reuters's controversial CLEAR database, as well as a commercial database used by LexisNexis, to deport a man who said to the immigration judge, "I came back to be with my family, I'm sorry, that's all" (Rivlin-Nadler, 2019).

By looking at government contracts among academic databases, others—academics, lawyers, librarians, and researchers among them—have found out within the past few years that they too were networked into this state of data body milieu, that they too were part of the information dragnet that leads ICE to deportations (Lamdan, 2019). With every search of LexisNexis, with every attempt to demonstrate Elsevier materials to patrons, people are unknowingly tied into this ICE "dragnet."

Data body milieu is our/your social media data, the library's databases, and the DMV databases (to name a few) working together to build a new borderland for surveillance and deportations. It's that sense that your information is somehow linked to the massive surveillance project happening with ICE. It's the uncertainty of data privacy that Dreamers have when using their smartphones to check the always-shifting status of DACA.[2]

Data body milieu is a way to understand how and why the Latinx immigrant data body is at the center of technological innovation and development. The development of new borderland technology includes consumer appeal: What is developed in the borderlands and based on the Latinx immigrant body may become uniquely entwined with consumer products that are modified and put into future technology marketplaces, and vice versa. For example, the 3-D technology used for self-driving cars is piloted as a 3-D border wall. However, that company, Quanergy Systems, also continues to keep one foot in Silicon Valley. The improvements and changes made from the 3-D border wall can change the technology of the self-driving car, and vice versa.

2. The DREAM acts, both federal and California, are a legislative solution, whereas DACA was an executive order.

This state is both materially grounded and hyperimagined. These technologies that make up data borders have material context: reliance on infrastructure, such as hardware, software, information labor. But it is intentionally rendered invisible and holds an affect around the ephemeral. Companies, creators, and manufacturers have a stake in obscuring *who* and *what* makes the information technologies with which we engage every day. Kate Crawford calls these systems an "atlas of AI" that extends everywhere from mineral extractions to warehouses, to classification schemas, and there is a stake in rendering the many parts to these systems invisible. "This colonizing impulse centralizes power in the AI field: it determines how the world is measured and defined while simultaneously denying that this is an inherently political activity" (Crawford, 2021, p. 11). Thus, these data borders are materially located *and* ubiquitously invisible. Media Studies scholar Camilla Fojas finds that borders are a site of cultural production. Fojas names the rising state of borderland surveillance optics as 'borderveillance': the integration of infrastructure and cultural symbols that make the border "an archaeologically layered visual space of policing within a technological matrix . . ." (2021, p. 3).

Invisibility and hypervisibility frame Latinx immigrant experiences in the United States. Latinx immigrants are often made hypervisible when called out for political motivations. Undocumented people are used as political fuel to encourage further xenophobia that can lead to ultranationalism. We also see the circumstances in which Latinx immigrants, the undocumented, or migrant workers' labor conditions have circulated through social media as romanticized or victimized, for political leverage. During the COVID-19 pandemic restrictions, immigrants were hailed for their hard work as essential workers despite the dangers they continued to face, especially in large group settings such as meat factories or in other food supply chain settings where COVID-19 was spreading (Jawetz, 2020). Increasingly during the climate crisis, social media memes circulate that show migrant farm workers surrounded by smoke from Western fires, still in the fields bending over and picking. In recent years the conditions of immigrant detention centers are visualized as masses of Latinx immigrants behind fences. The hypervisible image of Latinx immigrant labor is used for political motivating action. Undocumented immigrants are constrained by limited rights in the United States, including voting, access to many public benefits, and educational resources, to name a few, although they pay into those systems through several tax streams. The hypervisibility and invisibility of Latinx immigrants in the United States serve to augment the borderless data border state in which we find ourselves: The hypervisibility of Latinx immigrants and surrounding xenophobia motivates the development of borderland technologies, and the invisibility of Latinx immigrants is what that technology is designed to uncover.

A borderless data border centers immigrants and their data as the object of and requirement for surveillance-focused algorithmic technologies; yet those technologies have expanded and encroached upon the lives of everyone in the United States (and beyond). Although this data borderland and data collection seem abstract, all people are intimately drawn together as our data intermixes, intermingles, and is used to classify, sort, rank, and rate us. Some of us are divided through necropolitics—the policing of some through death—others are drawn together with further life chances, which is biopolitics.[3]

Latinx immigrants are valued as a data body, one that is used for technological design and is valuable as a source of data in and of itself. We see Silicon Valley physically reshaping around the US-Mexico borderlands. And US citizens are engaged in a constant state of borderland surveillance, to the point that they are intimately entangled with undocumented data surveillance.

Is my information being correlated right now to deport someone I may or may not know?

Does my social media data aid in detention and deportation in my hometown?

Am I contributing to the ICE technology network? The swab samples collected from immigrants at all detention centers and at the border are taken to a laboratory 78.4 miles from my office in Rhode Island. It is 81.2 miles from my family home in Murrieta to the border at Tijuana, and six miles from my family home to a detention center cited for the poor conditions of immigrant detainees (Esquivel, 2019).

"They're still using those space blankets?" A loved one asks this in response to images of immigrant detention centers showing children wrapped in foil swaddles, during one of the national ebbs and flows of paying attention to immigrant rights, adding "They used those when I was detained in the eighties."

When I search LexisNexis or use Amazon, there are visceral images that come to mind. Sights. Smells. I can feel the borderlands; I can see friends and family that have been detained or deported.

Can you?

3 An anonymous reviewer helped me to work through this language and conceptualization, for which I am grateful. While I can't cite that individual because of the blind review process, I do want to acknowledge their help in shaping these concepts.

CHAPTER 1

The Physical Borderlands, the Data Borderland

This time we made it. We crossed at night, it was cold as hell, I don't remember what month. I remember the big lights, big poles with bright lights at night. You could hear the quads. They had big trucks, we were hiding somewhere, you would see them pass you, and then you run and hide in another spot. At that time, I remember we had to run. The second time we moved we crawled like little soldiers. You pause, you must pass inspections. There were times we waited a long time, sitting down in the bushes. It was dark, cold, you look up and just see stars, clear, and it was beautiful.

—Juana[1]

PROPHECIES OF THE BORDERLANDS TO COME

Mexican/Chicano performance artist Guillermo Gómez-Peña wrote in 2001 about an emerging border: a virtual barrio. He observed that the discourse surrounding cyberspace and the Net of the 1990s and following decade was like that of the Western frontier: uninhabited, new, limitless, a "territory" where users could make themselves their own person, where everyone was "equal." The terms *originality* and *innovation* were commonly used for those creating virtual and digital art (2001, p. 197). On one side of that virtual border, he saw, were innovators, experimenters, creators. On the other side of that digital frontier

1. Interviews were conducted in English and Spanish. Those conducted in Spanish were translated and transcribed into English. After interviews were translated and transcribed, they were sent back to interview collaborators and their families for approval of the transcription and translation.

were Latinxs/Chicanos, African Americans, and Native Americans, discursively positioned as nontechnological, manual people.

With the public's increasing use of the internet in the 1990s and the building of the physical border and technological surveillance thereafter, Gómez-Peña envisioned a virtual borderland that would reproduce the same inequities happening in everyday lives. He believed that the fear and panic around Latinx immigrants and Latinx culture would also be transmitted across the internet. And he was prophetically correct. Rather than a virtual reality frontier, where inequality is applied online, data borders are like augmented reality, with the networked surveillance technology gathering data from every corner of physical and virtual life. The geographic borderlands, Latinx immigrant data bodies, and citizens' data make up the data grid for which to potentially surveil and deport undocumented people, and they are the landscape for which Silicon Valley's innovation can continue to expand and thrive.

When I spoke to people who had crossed the border without papers, they associated emerging borderland technologies with danger. The more surveillance technologies along the border, the more dangerous steps people crossing would have to take to evade them. Thirty-nine-year-old Juana, who had come when she was younger and lives in Lake Elsinore, California, discussed often how the emerging technologies were making the border more dangerous, a sentiment that is antithetical to the politicians who support the smart wall. As Juana put it: "Right now, I think they are using a lot of technology for everything. . . . A man told me you must cover your own footprints, because of the technology of tracing the footprints. The cameras, it seems impossible to come here, I mean it's not. Make it safer for people crossing in general so it's not about them being watched." In my interviews, developing information technologies and data tracking on the border were directly correlated with danger. Awareness of increasing technologies on the border among those who are undocumented, DACA recipients, or naturalized became a warning of the dangers of the borderlands.

We arrived in the current state of data border surveillance through a series of technological changes, the datafication of citizenship inclusion, and a sharp turn in economic value of bytes of information as a currency. Embedded racism, sexism, gender inequality, and classism are built into those systems; our data bodies replicate and exacerbate the physical world's stratified systems—even if we can't always visibly see where they exist. Especially because we can't always see where they exist.

In this chapter I review the history of the US-Mexico borderlands as they emerged geographically and politically, and then the recent move to introduce Silicon Valley's borderland surveillance project that expands the borderland as a data project across the country. I focus here on Mexicans, Mexican Americans, and Central Americans, because they have intensely experienced the borderlands as a place of migration, discrimination, and home. This is not to say that other immigrants and migrants do not experience the US-Mexico border or its surveillance technologies, but my focus here is on the populations that have resided and passed through these spaces most heavily.

The technology to surveil and capture immigrants over the years in the United States is often developed in response to the relatively challenging terrain conditions of the West. Information technologies are positioned as the bridge that can help the US state *finally* stop immigration through the US-Mexico border. Each new technology is sold as the solution that will manage the excess of immigration and immigrants, including Latinx immigrants who are unnaturalized or unassimilated. With each new investment into developing technologies, tech companies and state entities present the next hope that information technologies will bring control over Latinidad immigrant excess. One of the emerging themes in this developing data border is that the US-Mexico border, and anxieties about Latinx immigrants along and crossing through that border, are used as a launching point for innovation. The specter of Latinx immigrants drives the zeal of progress by way of technology.

A leading American value in the data border is that technology is a means of progress and that technology equals state futurity. Joel Dinerstein makes the case that "technology is the American theology" (2006, p. 569), in that it gives Americans faith in the future, "both in the future as a better world and as one in which the United States bestrides the globe as colossus" (p. 569). Technology is and has historically been hailed as the foundation of American progress, what he calls the technocultural matrix: progress, religion, whiteness, modernity, masculinity, and futurity (p. 571). Those underlying values shape technological design. Religion, in this way, is a faith that Americans have that technology will lead to advancement. They are the motivators for which we (technology users, designers, and Americans) put our faith in technology to solve a problem. Dinerstein concludes that new technologies "help maintain two crucial Euro-American myths: (1) the myth of progress and (2) the myth of white, Western superiority" (p. 572).

This American mythos invests in futurity by way of technology as well, especially in Big Data and AI. The development of borderland technologies, then, is a promise of futurity with the "right" kind of country of immigrants.

To understand how the borderland is becoming the next landscape in which technological innovation is executed, I first want to establish the history of the development of the US-Mexico border as place and space.

BORDER TECHNOLOGIES: A TIMELINE

The Developing Borderland

The US-Mexico border is a geographic place and a boundary in flux. Traffic between both countries makes for dense sites of interaction among residents. Much of the border as it is constructed today launched with the Treaty of Guadalupe Hidalgo in 1848. Mexico had become an independent nation from Spain in 1821, and the United States attempted to purchase Texas, New Mexico, and California without success. In the 1820s and 1830s Americans migrated heavily to Texas, which led to the annexation of Texas in 1845. The Mexican-American War of 1846 led to new boundaries of the borderlands with the Treaty of Guadalupe Hidalgo and the subsequent Gadsden Treaty, which drew the border to include New Mexico and Arizona as we know them today (Martínez, 1994, p. 32). Mexicans in the territories were given the rights to stay in the United States or leave for Mexico. Those who stayed could become US citizens or remain Mexican citizens with property rights (Daniels, 2002, p. 314). After the war, there were eighty thousand Mexicans in the United States, primarily in New Mexico, California, and Texas.

Mexico and the USA developed a businesslike international relationship from the 1880s to 1910, when the start of the Mexican Revolution led to more chaos and turbulent borderlands. After 1910, climate in the borderlands remained volatile, but commerce and economic growth continued. Anglo Americans moved south for investment opportunities upon the encouragement of Mexican President Porfirio Díaz. They were especially interested in states that had rich natural resources, such as Sonora and Chihuahua (Martínez, 1994, p. 35). Railroads played an especially important role in shipping crops and livestock into major urban areas. In the 1920s, the United States and Great Britain were welcome in Mexico to mine for oil. However, after large disputes Mexico nationalized the oil industry in 1938, and President Roosevelt acknowledged Mexico's right to "expropriate property belonging to foreigners" (p. 37).

Since the 1920, there has been rising concern over undocumented migration on the US side of the border. Nativism arose with phases of national anti-immigrant movements, toward various populations depending on the time (Daniels, 2002). The first phase was anti–Irish Catholic, from the 1830s to the mid 1850s. The second phase of nativism and anti-immigrant movements was anti-Asian, particularly Chinese immigrants, with the apex leading to the Chinese Exclusion Act of 1882, followed by anti- Japanese movements from 1905 to 1924, and an anti-Filipino movement in the 1920s and 1930s. The anti-Asian sentiment has manifested in such forms as the Japanese American internment camps of 1942. The third phase manifested as an anti-immigrant movement beginning in the mid 1880s, which led to the Immigration Act of 1924: "Successful nativist movements have almost always been linked to more general fears or uneasiness in American society" (p. 265).

Anti-Mexican movements began with the California Gold Rush, when competition increased among Mexicans, Chinese, and Americans. Mob violence and lynching was specifically used against Mexicans and Chinese miners in California (Daniels, 2002, p. 314). In the 1930s, Mexicans were already living in the states that were previously part of Mexico. They were not immigrants, but longtime residents of the Southwest. In the 1940s, the derogatory term for Mexican Americans was greasers, who were targeted by Los Angeles law enforcement and the Texas Rangers. Through the 1970s and 1980s, the Mexican population continued to increase; at that time, Mexicans and Mexican Americans were overwhelmingly farmworkers.

The English Only political movement formed in the 1980s in response to increasing populations from Mexico and increasing presence in Spanish-language media, including newspapers, magazines, radio, and television (Daniels, 2002, p. 318). Also during the eighties was the rise of the sanctuary movement, which gained momentum with the influx of Central American refugees, migrants, and undocumented immigrants and the US government's targeting of those groups. Sanctuary communities housed undocumented people targeted by law enforcement and the US government, especially Arizona, New Mexico, and Los Angeles (p. 384). It was during the seventies and eighties that surveillance technology projects received heavier federal investment.

Border Technology and Automating Race

The recent move to build a "smart" border wall and use biometric analytics is a revival of the past, not a new idea. Rather, it is an idea that is reintroduced during apexes of immigrant anxiety and xenophobia. The smart wall, through different technologies and military-commercial partnerships, reappears every few years with new energy behind it.

The same narrative of progress that moved railroads and telephone lines from the East Coast to the West Coast after the Mexican-American War moves Silicon Valley innovation along the US-Mexico border. These are the underlying cultural values that justify a heightened state of surveillance and the rationale that allows for such moves by the Department of Homeland Security as collecting DNA from immigrants. As Silicon Valley companies merge with defense companies and the DHS, they justify themselves with this American cultural value: innovation through technological progress moves our society forward in a promise of futurity and security. When we see language such as *innovative,* we know it is tapping into a value that lies deep in the American historical psyche.

In the seventies, the Immigration and Naturalization Service (INS) began using intrusion detection systems along the US-Mexico border. This technology greeted Mexican immigrants along the border as "data producers" (Chaar-López, 2019, p. 496). Cybernetics such as computers, ground sensors, and other technologies built what was considered the first electronic fence; the US-Mexico border was "one of the critical spaces where government experimented with automating the control of racialized populations" in the seventies (p. 498). Cybernetics began to flourish at MIT in the wake of World War II. During the Vietnam War, the US military developed the "McNamara Wall' or "McNamara Line," a system of integrated computers that detected enemy intruders that began with experimentation in South Vietnam but was transferred over to the US-Mexico Border (p. 506). Automation was justified along the border by way of the race and racism toward Mexicans, building up through the sixties—"'the border,' an assemblage of artifacts and practices, has been a technology designed to administer racial inclusion and exclusion" (p. 515). Surveillance technologies were increasingly designed around the Mexican body as the object of capture.

In the mid-eighties, immigration along the border increased, and in 1993 the United States began constructing fences along the San Diego–Tijuana border, increasing border patrol through Operation Gatekeeper

(Etter and Weise, 2018). Immigrants crossing began to die along the border in increasing numbers, and by the start of the twenty-first century there were two hundred migration-related deaths per year (Etter and Weise, 2018).

From 1997 to 2005, the US government spent $429 million on technology programs that are considered to have been unsuccessful (Davis, 2019). With the terrorist attacks of September 11, 2001, there was increasing focus on the border as a possible port of entry for terrorists from around the world. Congress passed the bipartisan Secure Fence Act of 2006 to build seven hundred miles of fence (Etter and Weise, 2018).

2006: The Secure Border Initiative

The virtual wall began in 2005 through DHS's Secure Border Initiative (SBI), negotiated as a $7 billion deal with Boeing to build SBINet for the estimated two thousand miles of border (Preston, 2011). SBINet, a series of sensors, radars, and cameras on towers, is still used by border patrol to this day. It led to a total of nine towers along a twenty-eight-mile stretch of Arizona, "with each tower including surveillance and target-acquisition radar arrays and an electrooptical/infrared camera. Between the towers, ground sensors would be embedded out of site, relaying information back to that network" (Bump, 2016).

After a total expense of $1 billion, the project was halted by Secretary of Homeland Security Janet Napolitano in 2011 because it was considered poorly managed and much of the technology was defective (Davis, 2019). Foreshadowing today's proposals, "the original concept of the project, to develop a single technology that could be used across the entire border, was not viable. Boeing had built a complex system of sensors, radars and cameras mounted on towers that was supposed to lead border agents to the exact location of illegal crossers. 'There is no one-size-fits-all solution to meet our border technology needs,'" said Napolitano (Preston, 2011).

What came from SBInet was a series of integrated fixed towers and cameras with motion sensor systems. Increasingly, government officials advocated for technologies specific to the different geographic conditions of the border: "The setting up of an electronic fence in the Sonoran Arizona desert is far different than doing so around San Diego's urban area," said Alan Bersin, border czar and US Customs and Border Protection commissioner from 2010 to 2011 (Davis, 2019).

Coinciding with the Boeing contract were further systems of surveillance and mined data surveillance. The Department of Defense's Total Information Awareness began mining transactional data to predict terrorist activity. The smart border was built through the Secure Electronic Network for Travelers' Rapid Inspection system (at the Tijuana–San Diego border) and Nexus, the US-Canada airports and border surveillance program. "Virtual borders of the United States would extend to the point of origination for visitors. The bulk of the security checks will be performed at the time of application for a visa to visit the United States (Root, 2004)" (Puar, 2007, p. 155).

With the expiration of the Boeing contract, many law enforcement officials and Republicans criticized President Barack Obama for what they considered a lax approach to border enforcement, perhaps also foreshadowing the sharp increased investment in border technology by the Trump administration. After the SBInet failure, millions of dollars were put into more surveillance towers, cameras, drones, and thermal imaging devices. But those systems are now considered aging and flawed, the sensors tripped frequently by animals, weather, and border agents themselves (Davis, 2019).

2015–17: The New Data Borderland

More recently we have seen a shift in the technology and startup industry as Silicon Valley relocates to the geographic southwestern borderlands to build a digital border wall. In 2015, DHS opened a satellite office in Silicon Valley. The language around that partnership would suggest that a new era of American innovation would emerge, with fresh new ideas and funds to invest in these new technologies:

> In April 2015, the Secretary for the Department of Homeland Security (DHS) announced DHS would establish a satellite Silicon Valley Office (SVO) to engage entrepreneurs and innovators from small startups to large companies, incubators, and accelerators. The SVO's mission is to tap into the innovation of the private sector in new ways, opening the doors for nontraditional performers to work with government and to pose the government as a viable customer for their technology. (Department of Homeland Security [DHS], 2015)

Innovation and *innovators* are motivating terms in this relationship between Silicon Valley and DHS. This, of course, was not solely about the US-Mexico borderlands; the net was cast over the entirety of homeland security: "While DHS will be seeking solutions to challenges that could range across the entire spectrum of the homeland security mission, the first challenge will be in the area of the Internet-Of-Things Security" (DHS, 2015).

As DHS collaborated with Silicon Valley in 2015, the promise, the excitement, and the American spirit were all present in the language of this partnership that aligns with the Western values of ingenuity, innovation, and progress. The message was that Silicon Valley could now save America from homeland threats. This process is reflected in what DHS called "Ideation workshops": "Through the ideation process, DHS will educate tech companies (from startups to large market-shaping companies to accelerators/incubators) on DHS's need for better security, safety and resilience in the capabilities we employ and deploy" (DHS, 2015). Technological inventors such as Alexander Graham Bell, Nikola Tesla, Steve Jobs, and Mark Zuckerberg are invoked in this spirit of innovation. The vision of their workshops and garages filled with computers is present as DHS calls for those workshops as a place to tinker for "full creativity to develop solutions" (DHS, 2015).

In 2016, President Donald Trump won the presidential election riding on the long-planted vitriol among some American voters (enough to carry the electoral vote) toward Mexico and Mexicans (among many other xenophobias and racisms). He stated at the beginning of his campaign in 2015 about Mexico: "They are not our friend, believe me. . . . They're bringing drugs. They're bringing crime. They're rapists. And some, I assume, are good people" (Phillips, 2017). He also named Mexico as an enemy of the United States. In that same speech, he said: "When Mexico sends its people, they're not sending their best . . . they're not sending you. They're not sending you" (Phillips, 2017).

The border was used a site of visual stimuli for voters: "You look at countries like Mexico, where they're killing us on the border, absolutely destroying us on the border" (Reilly, 2016). After Trump took the presidency, Democrats gained control of the House of Representatives; the border wall funding, proposed to come from taxpayers, became a fight that threatened to shut down Congress.

In 2017, the Government Accountability Office (GAO, 2017) conducted an audit of technology used by border patrol along the US-Mexico border . Table 1 shows the technologies installed throughout California, Arizona, and Texas at the time of that report. These technologies have developed over the decades in relationship to Boeing's failed SBInet wall and through increased funding and troubleshooting of surveillance along

TABLE 1 INFORMATION TECHNOLOGY ALONG THE BORDER, CONTRACTED BY THE US GOVERNMENT

Technology	Location	Description
Agent Portable Surveillance System (APSS)	Arizona, Texas, and California	Radar, daylight, and infrared cameras and a laser illuminator. Portable, does not link to a command and control center.
Integrated Fixed Towers (IFT)	Arizona	Radar, daylight, and infrared cameras mounted on fixed towers. Linked to a command and control center.
Mobile Surveillance Capability (MSC)	Arizona, Texas, and California	Radar, daylight, and infrared cameras; a laser range finder; and a laser illuminator mounted to a truck. Information linked to the truck information center.
Mobile Video Surveillance Systems (MVSS)	Arizona, Texas	System of live infrared cameras, laser range finder, and a laser illuminator. Control station in the cab of the truck.
Remote Video Surveillance System (RVSS)	Arizona, Texas	Daylight and infrared cameras and a laser illuminator linked to command and control center.
Relocatable RVSS	Texas	Daylight and infrared cameras and a laser illuminator, mounted on an eighty-foot tower, linked to a modular command and control center.
Thermal Imaging Device (TID)	Entire southwestern border	Portable, handheld infrared camera and viewing kit that can see up to five miles in the dark. Not linked to command and control center.
Unattended Ground Sensors (UGS) and Imaging Sensors (I-UGS)	Entire southwestern border	Ground sensors and cameras. Communication equipment sends information to the command and control center and to border patrol agents.

the border. Borderland technologies of the past act as a network that Silicon Valley will link into with more advanced systems.

2018: The Virtual/Smart Wall Is Funded

In 2018, Congress approved the $1.375 billion budget for 1,954 miles of border wall; $400 million of that was guaranteed to go toward the smart wall (Davis, 2019). There is a shift from defense companies such as Boeing and Elbit, still building Integrated Fixed Towers along the border, to a bid by Silicon Valley startups to develop newer information technologies around Latinx immigrant data, as well as network into that technology that is already developed or developing. By 2023, Truthout estimates that data gathering on the US-Mexico border will be a $740 billion industry (Chen, 2019), creating an entire borderlands industry as a new Silicon Valley.

As we enter this new era where Silicon Valley's motivation is to collect immigrant data, the larger narrative here is the promise of futurity through technology. This promise leaves out Latinx immigrants, or includes only the right type of Latinx immigrants. For example, recently designed Latina AI virtual chat agents, such as USCIS's now defunct chatbot "Emma," are designed with olive skin tone, speak English first and then Spanish, and fit the ideal Latina image preferred in major media in the past (Sweeney and Villa-Nicholas, 2022). AI designed to be Latina demonstrate the "right" kind of immigrant that is increasingly quantified: designed by tech companies if they can't be assimilated from within immigrant groups.

The Silicon Valley Borderland

To make bids for the digital border wall, Silicon Valley startups and tech giants now physically locate themselves closer to the US-Mexico border, using southwestern deserts as a test space for their innovation labs and Latinx immigrants as their test subjects. This new era of innovation casts the old defense test spaces such as Bell and Sandia labs with new Silicon Valley legends around building a transformative technology in a garage or home. The southwestern border invokes that energy of potential for hackers, makers, and tinkering spaces.

Visual images of Latinx immigrants, sometimes in large crowds, sometimes as sole silhouettes crossing an arid landscape, often preface the demonstration of a new technology that could be "the one" to stop people from crossing the border. In a CNBC report (2018), Silicon Valley companies demonstrated the virtual border wall. The failure of the physical wall and diverse southwestern landscapes is always present in these narratives: a large wall will not end immigration. The physical wall is viewed as a burden for those residents who live in Texas and do not want to view a physical structure from their homes. Latinx immigrants are portrayed as drug smugglers, criminals, or people looking for opportunity.

Silicon Valley becomes the answer to these many posited problems; the companies' physical branches in the Southwest are pitched as a more humane answer to the physical borderlands of the Southwest. One example of this is Quanergy Electric, which had been sending employees down to the Southwest border since 2017 to test new technologies. Quanergy pitched a three-dimensional wall built of LiDAR sensors, the technology that has been used for self-driving cars. LiDAR can detect the differences between animals and humans, is resilient in adverse weather, and improves nighttime vision. They proposed to create a two-mile perimeter. These companies propose an invisible response to the physical problems experienced with past technology in geographically wild and remote locations.

Known Trump supporter Palmer Luckey's Anduril Industries, headquartered in Orange County. frequently tests at the Southern California border. At a tech conference in 2019, Luckey described an established experiment lab at the border as he pitched for border wall funds with a proposal to build lattice software systems, machine learning AI: "What we're working on is taking data from lots of different sensors, putting it into an AI-powered sensor fusion platform so that you can build a perfect 3D model of everything that's going on in a large area. . . . Then we take that data and run predictive analytics on it, and tag everything with metadata, find what's relevant, then push it to people who are out in the field" (Fang, 2019).

Anduril has won contracts with various parts of the US military, including building a perimeter surveillance system for Yuma, Arizona, near the southern border. Perhaps what is most striking about Anduril is that it is in Southern California, where many other defense companies are located, such as Lockheed Martin and Northrup Grumman, instead of the traditional Northern California or Seattle-based locations of many "Silicon Valley" tech companies.[2]

2. "Silicon Valley" serves as a place-based name for the nexus of information and computing technology industries in the Santa Clara Valley, as well as a symbolic name for global technology cultures and networks. Silicon Valley has an original geographic location that has gentrified and imposed cultural values in Northern California, but also circulates culture and technological products, and migrates companies beyond that geography, for example, the expansion into Los Angeles, "'rebranded as 'Silicon Beach'" (Noble & Roberts, 2019). This expanded view of Silicon Valley allows for tracing how the material and cultural landscapes shape, and are shaped by, local and global networks, structures, labor, identity, ideology, and politics.

In the 2010s, the geographic US-Mexico border became the new Silicon Valley test lab.

The Expanding Borderlands

The data border is a bit different from the virtual/smart wall but not unrelated. The data border consists of all the ways in which the US-Mexico borderland is spread out around the United States, both digitally and physically by way of technological anchoring. This includes US citizens and permanent residents who are included in border patrol–level surveillance tactics.

Most striking about these emerging surveillance consumer technologies is that while they are tested and located at the physical US-Mexico border, and justified based on the Latinx immigrant threat, they are applicable to two-thirds of the population of the United States, including citizens. The expanding presence of ICE and CBP extends the surveillance state into many towns and cities across the country, with the specter of the Latinx immigrant threat always present as justification for random stops and searches. According to the American Civil Liberties Union (ACLU), the federal government considers ports of entry to be places where routine stops and searches are permitted. Because federal regulations give CBP authority to operate within one hundred miles of the US external boundary, CBP can operate immigration checkpoints anywhere in that zone, one hundred miles from all borders including the ocean, which impacts about two hundred million people (American Civil Liberties Union [ACLU], n.d.).

Engagements with border patrol have been normalized for quite some time for those who grew up close to the border. In 2017, ACLU obtained emails that found that the Obama administration granted CBP more access and fewer restrictions against openly searching transportation modes such as buses and trains (Kaplan and Swales, 2019). In 2019, passengers of Greyhound buses around the country experienced a new border surveillance state. Mercedes Phelan, who is black and Puerto Rican, experienced interrogation from border patrol on a Greyhound bus in Pennsylvania and an Amtrak in New York in two different instances in the same year. She recalled, "They literally skipped over every single white person" (Kaplan and Swales, 2019). Reports of border patrol stopping buses close to the Canadian border are more common, but also expanded with the Trump administration, and they are known for their racial profiling practices. When defending these new search procedures, CBP invokes the image of the immigrant smuggling illegal drugs as the threat and justification. They say they are searching for "alien smuggling and drug trafficking organizations to move people, narcotics and contraband to interior destinations" (Kaplan and Swales, 2019).

If technology, then, is the mover of our society toward futurity, borderland technology becomes a promise to cure American states of unease. The Latino threat, as Leo Chavez (2013) has named, has been cast for decades around the nonassimilating immigrant. The hype around merging Silicon Valley with defense industries to build border technology becomes part of the promise of a future without Latinx immigrants.

The continued justification for borderlands technology as a mover for the greater good of society is a phenomenon that Ruha Benjamin (2019) calls the New Jim Code: the timeline of technologies becomes natural for catching bad guys, and the refinement of those technologies promises to help manage the Western terrain and separate the accepted forms of Latinx immigration from the unwanted. Dinerstein (2006) observes, "Technological progress is the quasi-religious myth of a desacralized industrial civilization; it is sustained through new technological products, not empirical social change" (p. 591). Amid the debate over the budget funding, proponents advocated for the smart virtual wall as a more humane version of the physical wall. The technology proposed for the virtual border has been described as "non-intrusive inspection technology" by Senator Dick Durbin (D-Illinois), intended solely to look for narcotics trafficked from Mexico into the United States (Durbin, 2019). In contrast to the physical wall, the virtual/smart wall was pitched as cheaper and more effective, welcoming in the smart wall with a smoother transition than the physical border wall.

The physical and the digital data border is a commercial product so valuable that tech companies reorganize around the capturing and collection of that data, which can be networked into multiple systems of surveillance, mined by AI, and produce new information systems. That the Latinx immigrants as a source of data would become the next concept to "like," "heart," and "click." A data currency. That Latinx immigrant data would be the new frontier.[3]

3. To make this data border more visible, if you, the reader, are in the United States, you can consult the news outlet Sludge's US CBP vendor map, compiled by Alex Hotch: https://public.tableau.com/app/profile/alexkotch/viz/CBPVendors/Dashboard1. This database enables searchers to look up their geographic proximity from vendors that collaborate with CBP.

As the US-Mexico data borderland expands throughout the United States, Latinx immigrant data bodies and biological data is also located around the United States by way of biometric information: face scans, fingerprints, tattoo scans, retina scans, full-body imaging, and increasingly DNA are located physically and in the cloud without one solid geographic location. While Latinx immigrants settle throughout the United States to make new homes beyond the US-Mexico border, and the borderland milieu may exist as far away as South Carolina, what has shifted is the Latinx immigrant's biometric information removed from that person and distributed throughout various data storage facilities across the United States.

I explore the Latinx immigrant data body further in chapter 2, along with how it becomes the situation for which Silicon Valley companies are now designing border surveillance and consumable technologies that may be next year's hot Black Friday gift.

CHAPTER 2

Latinx Data Bodies

I remember the border patrol had their headlights on us on their quads [four-wheelers], there was border patrol on horses. They surrounded us and we gave up and the next thing I remember was that they had us surrounded. The detention center was cold, and I asked my mom if they were going to kill us, because the whole time they were talking to us the border patrol had his hand on his gun, so I was terrified for my sister and my mom.

—Oscar

Immigrants' rights activists and immigrant law attorneys have noticed a recent change in deportations. For decades, deportations were done en masse at workplaces known or suspected to hire undocumented people. However, in the early 2010s, ICE focused on deportations at locations that seemed impossible to know—at courthouses before a court date, in homes, in front of workplaces (Bedoya, 2020). ICE agents would arrive in places that were intimate and individualized. In the 2010s, it was publicly known that algorithms were determining who and when to deport, and not only in the United States. In the United Kingdom in 2016, thirty-six thousand students had their visas revoked by a computer program that claimed they cheated on their language tests. These automatic deportations were wrong in about 20 percent of the accusations, "meaning that more than 7,000 students were likely to have been wrongly accused of cheating" (Baynes, 2019). In Canada, algorithms, and AI "augment human decision-making" for Canada's immigration and refugee system (Molner and Muscati, 2018). The turn toward big data and AI is worldwide, but we will keep our focus on the US-Mexico border and deportations local to the emerging digital borderlands in the United States.

Latinx immigrants in the United States have long been an object of politically discursive propulsion for surveillance projects and military agendas. Fear, anxiety, and xenophobia are motivators for investment in surveillance technologies. At a time when there is active bidding for contracts for this approved smart wall, there are many mockups on how tech companies are going to collect immigrant data in the borderlands. The smart wall goes beyond what would seemingly be the purpose of immigration policy—to stop people from coming into the United States and to apprehend and deport immigrants in the borderlands.

The combination of Silicon Valley and DHS has led to Latinx immigrants being considered data. Moreover, that data has evolved quickly from the information along the border to data networked into digital technologies such as social media accounts and academic databases, and further, to Latinx immigrants' increasingly valued biodata. Kate Crawford notes that the last decade has dramatically seen the capture of material for AI production: "This data is the basis for sense-making in AI, not as a classical representation of the world with individual meaning, but as a mass collection of data for machine abstractions and operations" (2021, p. 95). Part of the nuts and bolts of algorithm building is training data sets. "Training datasets, then, are at the core of how most machine learning systems makes inferences" (p. 97). Inherent in these values of building machine learning is that more data equals better improvements to information systems, which equals better data in turn. Capturing data en masse is the avenue to improving and evolving technology (p. 112); lack of consent by the people involved is the premise for building borderland technology. The more immigrant data, the more data to improve those systems, and the more inferences made because of that data. In computer sciences, information science, and many STEM fields, this process of mass gathering of data without consent is presented as a benign act because it leads to the improvement of other technologies. Therefore, gathering immigrant data as a product begins at nonconsent. Latinx immigrant data in borderlands is caught up in this religiously held belief system that technological improvement is equal to futurity and the progress of Western society.

Many of us have experienced the value of our data when using technology. For example, when you shop online, scan through your Facebook feed, or an ad for a product suddenly pops up while you're scrolling through a website just after you said out loud to a friend that you were thinking of buying it, this is the high value of your data at work. An algorithm, or many algorithms, determines which information you will engage with next.

The information product sold, the information building AI algorithms, and the test subject for new technologies without permission—all are immigrant data. As undocumented people, they are often outside of the scope of ethics that shields citizens and (some) permanent residents from exploitive data collection. The zeal for technological innovation reflects the spirit of American competition for immigrant data as the information product to design around. The coding of these technologies is built around the gaze of the Latinx undocumented person as the object to capture or the desired data. This drive for Latinx immigrants' datafication at and around the border relies on narratives saying that Latinxs have previously had little to no history in the development of the histories of technology, or that Latinxs are biologically suited for manual labor in technology industries.

This chapter explores how Latinx immigrants have been and are becoming valued as data products. First I review how Latinx undocumented immigrants are conscious that they are surveilled through information technologies, then how Latinxs have been historically embedded in the technology industry, and finally how recent companies such as Quanergy, Coolfire Core, Anduril, and BI2 develop technologies around Latinx immigrant data as an information product and how those data bodies lead to new technologies designed for further surveillance projects. Woven into the underlying assumption of quantifying Latinx immigrants is the discourse that positions immigrants as nontechnological and as unwieldy as the Western frontier itself.

LIKE WE'RE BEING WATCHED

My interview collaborators were highly aware that they are watched and datafied through their technology. They use technology for everyday lives on their smartphones, with computers, and video games, and technology is often used as a supplement for benefits from citizenship. During the presidential administrations of Trump and Biden, the young people that I interviewed who had received DACA or were hoping to apply for it would use their phones to check their DACA status continually (in the data collection phase of this book, DACA had not been reinstated).

None of the people that I interviewed felt that their information was private through their technology, "It's kind of scary," twenty-four-year-old Teresa, who has had DACA status since the Obama administration, told me, "I know if I search 'Sushi places near me' sushi stuff will pop up in other apps. It makes you wonder what else. . . . I feel like, if they really wanted to, they would know where I am."

One of the most insecure parts of being a DACA recipient is that young people had to provide all the information about how and when they crossed the border, every address they ever lived at, and the names of all of their family members. For Teresa, she had to exchange her information and further surveillance for more citizenship benefits: "[When DACA started] we really didn't know how legit this would be, if there are going to be repercussions, they know where you work, everything, it is scary."

My interviewed collaborator Juana described a phenomenon that many people experience around surveillance technologies. She told me that when she visited her sister-in-law in El Centro, she wondered about heat and body sensors reading her anxiety as they drove through border checkpoints. She had heard from friends that were in detention centers that they saw these heat sensor reading technologies used by center officers. A sense of being watched, while not knowing when or where one's data is being used, is a common thread among Latinx immigrants.

This is to say that Latinx immigrant communities are aware that they are datafied and surveilled through many avenues. Many people exchange stories with those who have been in immigrant detention centers about the types of technologies used. As further discussed in Part 2, Latinx immigrants weigh the costs and benefits of exchanging their data for entrance into the privileges of citizenship.

THE QUANTIFIED SELF

In recent years, the "quantified self" has emerged prominently through bio-tracking devices such as smart watches. Hacking self-improvement through biotechnology is sold as a way of augmented human potential. Quantifying Latinx immigrants at the US border and beyond is also a mode of the quantified self, encouraging residents to produce data that contributes to these surveillance networks and proclaiming that a good citizen is a datafied citizen.

This 'good citizen,' this quantified being, has developed over time in the United States. In the late nineteenth and the twentieth centuries, companies used statistics to predict and determine risk; in the late nineteenth century, insurance company's predictive analytics through data discriminated against African Americans in their insurance policies. Black civil rights activists demonstrated how insurers' datafied tools discriminated against Blacks and affected their likelihood to obtain insurance. Risk and predicting risk through data "became a kind of commodity" in the insurance field (Bouk, 2015, p. xx).

Sarah Igo writes that average American citizens became known and conscious of themselves as a populace by way of mid-twentieth-century surveying tools and a flurry of social scientists dispersing to find data on the "averaged" citizen. Public surveys led to a public consciousness about the statistical public (Igo, 2007). Quantifying the self as a form of state making is a project of knowing that is not necessarily based in universal truths: "We need to understand social scientific representations—of 'typical communities,' 'majority opinion,' and 'normal Americans'—not as reflections of the body politics but as an index to political and epistemological power" (Igo, 2007, p. 22).

The averaged, normative, quantified citizen in the United States has been a long-forming historical construct. However, the shift with quantification and race came in the twentieth century. Jacqueline Wernimont argues that what she calls *quantum media*—for example, census records—have long been quantifying human life for purposes of control and profit mostly for state growth; but in the twentieth century, quantifying people shifted: "quantum media historically have represented a small, privileged section of the population as persons valuable to the state, or after the twentieth century, as valuable to corporations and 'human knowledge.' Throughout the same time, nonwhite people have been refigured by quantum media as property, depersonalized data sets to be used as resources or liabilities rather than as people" (2018, p. 161).

Quantifying Latinx immigrants through data and biotechnology is a means of further control—to quantify an undocumented person invites the threat of no data or incorrect data (for example, undocumented people who use false social security numbers). Quantifying immigrants also allows the government to put those people back in their ordered place; removing them from the unpredictability of borderlands, markets, and industries in the United States. Sun-Ha Hong notes that this process of datafication results in the good, modern, neoliberal subject: "Datafication thus reprises the enduring dilemma around the modern ideal of the good liberal subject: individuals who think and know for themselves, their exercise of reason founded on free access to information and the possibility of processing it fairly" (2020, p. 3).

When American citizens are discussed in any manner in relation to the quantified self, there is always an undiscussed element of undocumented immigrants in relation to that community. Undocumented people erode this quantified self by a lack of social security numbers, false information, or moving through undocumented spaces such as the wild terrain of the United States. They must be datafied. While they do not qualify as the "good" neoliberal subject, these technologies draw them into a controlled fold of belonging by making them data points via this smart wall.

LATINX TECHNOLOGY HISTORIES

While Latinx people have always been present in the history of computing, archetypes around Latinx manual labor as a bridge toward technological development appear alongside active forms of agency, resistance, and collaboration with emerging technologies.

Immigrants along the US-Mexico border have long been valued as information sources. As early as the nineteenth century, Mexicans were regarded as biologically necessary to deliver emerging technologies and westward expansion. Track, maquiladora, and telephone workers exemplify *how* Latinx labor justified new technology. Restructuring the tech industry around Latinx immigrants is not an isolated contemporary event.

Railroad companies recruited Mexican migrant workers to build the new technology of the United States during 1870–1930, which this set a precedent for Latinx digital labor and emerging data value. Media sources labeled Mexican men as "semi-skilled" and "common labor" as they built the railroads in the Southwest and Central Plains. Labor agencies subcontracted track workers and the companies did not directly employ them, a pattern we see as well in Latinx technological history that emerged with neoliberalism and in the shifting labor landscape that Latinas in telecommunications have seen during their lifetimes.

Mexican and Mexican American railroad track workers were known in Spanish as *traqueros*. Railroad labor had previously been Irish, Black, and Chinese men in the 1800s. Mexican labor increased in the early nineteenth and twentieth centuries due to discrimination, the Chinese Exclusion Act, and the exclusion of Black men. Media outlets, railroad companies, and the labor subcontractors described traqueros with superhuman abilities. Their labor aligned to deliver manifest destiny by building the railroad (Garcilazo, 1995, p. 13).

With the traqueros, the message of American progress through Manifest Destiny justified the necessity of their labor and the biological strength of the Mexican man's body. Media outlets published excited proclamations around Mexican male labor that objectified them as superhuman to bring this new technology across the US and as an object, or a product, necessary for assembling the train's technology in the United States (McHenry County Historical Society and Museum, n.d.). A common theme throughout Latinx technological labor is this handoff from Asian labor to that of Mexicans and Latin Americans. Traqueros were desired resources after the Chinese Exclusion act. Mexican men were associated with the efficiency and super strength of the train itself, not confined to the Southwest. Those rhetorical characteristics acted as the reasoning to remove Chinese, Black, and Native American workers from the railroad.

That railroad infrastructure would later shape the laying of the fiber-optic network. "Telecommunications companies quickly recognized the value of rail right-of-way as real estate for running cable networks long before the Internet" (Burrington, 2015). The railroad led to the network for telegraph and telephones, which led to fiber-optic cables along rail routes, which grew during the first dot-com bubble. Current data centers, such as Google's Iowa data center, are near the railroads (Burrington, 2015). The Mexican body, justified as biologically suited for manufacturing labor, became means by which this technology arrived.

Latinas were more publicly associated with technology labor as factories moved out of US urban midwestern areas such as Detroit and into maquiladoras beginning in the 1960s.[1] The Radio Corporation of America moved down to the borderlands in the sixties, where Mexican women's labor would produce many electronics that circulate globally (Cowie, 2001). Maquiladoras sought to hire unmarried, female laborers, leading to gendered divisions of labor. Like traqueros and Chinese women's technological labor, Latinas were deemed biologically suited for electronics production (Cowie, 2001, p. 119; Nakamura, 2014). Mexican women, Chinese women, and Navajo women were cast as 'inherently and biologically flexible labor' with super-detailed capabilities for information technology labor (Nakamura, 2014, p. 929). Latinas at maquiladoras assemble a large percentage of electronics on the border, including computers, memory chips, network switches, routers, televisions, circuit boards, liquid crystal display (LCD) panels, mobile

1. Maquiladoras are factories built close to the US-Mexico border as a response to the liberalizing global market. As companies began to outsource labor into countries so that they could evade unionizing and living wages, maquiladoras were increasingly developed. See Devon G. Peña's *The Terror and the Machine: Technology, Work, Gender and Ecology on the U.S.-Mexico Border* (University of Texas Press, 1997) for more.

phones, communications equipment, and electronic appliances (North American Production Sharing, 2016).

This labor in technological infrastructure is not without agency and resistance to assumed ideologies around capitalism, technology, and progress. Mexican women's narratives of maquiladora labor are examples of agency and struggle to prevent such tedious technological work from dominating their lives. Agency and skilled knowledge go hand in hand. Latinas in maquiladoras acquired both mundane and highly skilled engineering techniques in technology assembly, using these skills as avenues of agency: "Workers, like managers, engineers, or social theorists, are perfectly capable of creating and articulating their discourses on moral rights and deconstructions of the will to power on the shop floor" (Peña, 1997, p. 184). Latina maquiladora workers that assemble electronics are critical about the products' harm toward the environment and the expectation of nonstop labor (Peña, 1997).

Latinxs also worked and still work heavily in telecommunications sectors. First recruited around early seventies with the 1973 AT&T-EEOC Consent Decree, Latinxs worked as telephone operators, linemen, data entry clerks, and in many other blue- and white-collar jobs in telephony. In telecommunications, the technology of the telephone acted as a buffer for Latinxs' nonwesternized, US-assimilated features. For example, telecommunications companies recruited Latinx telephone operators to speak Spanish at the operator's switchboard but prohibited them from speaking Spanish from the switchboard to the exit (Villa-Nicholas, 2022). Latinas were often denied promotion to more technologically sophisticated work, on the justification that their personalities did not fit in with technological labor beyond blue-collar work. Again, we see Latinx bodies and culture as bridges for delivering technology and deemed biologically unsuitable for advanced technological skills.

Latinxs have long implemented agency into their development and adaptation of information technologies by way of education, corresponding, and uniting with civil rights movements. Since the sixties, in information educational jobs like librarians, Latinxs in the United States had been creating information technologies not only to organize information but also to parallel Chicanx rights movement goals. Identity formations intertwined with information technology development in libraries. For example, Latinx librarians adopted multilingual software that resisted the English Only movements of the late eighties, such as the Toltran System Multi-Lingual Software, encouraged Latinx youth to develop multimedia materials through video recording in the nineties, and built computer literacy programming within their neighborhoods and library communities (Villa-Nicholas, 2015). In the nineties Latinx librarians indexed electronic resources that supported the southern Mexico Indigenous group and movement of the Zapatistas, advocated for more public access to internet and hardware, and argued that telecommunications access was a common public good and right (Erazo, 1995, p. 16). Forewarning of system bias through classification systems, Latinx librarians protested early forms of stereotyping classification; the library professional organization REFORMA organized a protest against PAIS, an image database that classified Latino gangs as representative of all Latinxs in general (Villa-Nicholas, 2015, p. 552). Humanities librarian María Teresa Márquez created the first Chicanx electronic mailing list in 1991, named CHICLE for Chicana/Chicano Literature Exchange. Borderlands historian and library scholar Miguel Juárez (2017), also a founder of an early Latino history email list, observed that "These electronic lists were influential in expanding communication and opportunities among Chicanx. . . . The Internet revolution was largely shepherded by librarians in their institutions." Latinx librarians developed and applied information technologies in public places, K-12, and academic libraries as incubators for technological innovation.

Latinxs have historically been embedded in technological design and even built the underlying infrastructure of surveillance networks. Latinxs were discursively situated as biologically suited for developing the train, electronics, and the telephone, but workers within those systems used their agency and critical frameworks to disrupt the presumed neutrality of those technologies (Peña, 1997).

IMMIGRANTS AS SUBJECTS OF INFORMATION

In the twentieth century, the growing data gathering techniques at the US-Mexico border coincided with the similarly growing set of techniques around what Sarah E. Igo calls the *averaged American.*

While census techniques for the purposes of taxes and building military by way of addresses have a long global history, the gathering of data around American citizenship shifted course in the mid-twentieth century for the purpose of *knowing* the *averaged American* (Igo, 2007). Social science tools such as polling and *quantifying* the national populace became a trend that held a mirror up to American citizens to reveal what the patterns of the public were; quantified data by way of surveys shifted the imagined community through numbers: "citizens could see themselves as part of a new collective, one constituted by and reflected in data compiled from anonymous others" (p. 6). Not only did social science techniques of gathering data take off at that time, but so did Americans' consciousness of themselves (Igo, 2007).

During the mid-twentieth century, shifting American consciousness around ideologies of what is normal paralleled an increase in data gathering of Latinx immigrants along the US-Mexico border (Chaar-Lopez, 2021; Ngai, 2014). Gathering Latinx immigrant data and converting that data into information is an established borderland practice. Border patrol has long tracked immigrant data throughout the borderlands by reading markers left by humans. These data points, or markers, include broken twigs, litter, and footprints (Lytle Hernández, 2010). Tracking along the border has a long history of racialization. One of the early tracking experts and tracking teachers in the border patrol was Fred D'Alibini. He organized tracking data racially: "A Mexican always walks heavy on the outside of his feet. When he walks, he puts his foot down on the heel first and then rolls it off—Indians will do that, too. Whites and blacks ordinarily put their feet down flat" (p. 49). Race, gender, complexion, and nationality were assigned to the data. Border Patrol Academy taught "sign cutting," the tracking of immigrants through signs. Chaar-López (2019) describes the process in two parts: "First, agents broke that large border terrain into discrete or manageable segments to scrutinize. And second, agents severed the path of those attempting entry-without-inspection; track-producing subjects would be removed from the border and their tracks would reach a dead end. Sign cutting strove to transform intrudes from unknowable entities into knowable, excludable subjects" (p. 503).

The first electronic fence of the sixties and seventies also organized Latinx immigrant data as a source of information. Like today's gold rush for data collection, storage, and retrieval, the Immigration and Naturalization Service (INS) viewed the borderlands "as a rich communication and information landscape with its particular set of practices (i.e., sign cutting) and sign-producing subject/objects" (Chaar-López, 2019, p. 503).

CORRELATIVE NORMATIVITY

Big data, predictive analytics, algorithms, artificial intelligence, and geocoding are powerful methods entrenched in our everyday lives. However, these technological tools are historically designed with bias and are not neutral or value-free.

The simplest definition of big data is from the Oxford Learner's Dictionary: "Looking at large amounts of information that has been collected on a computer and using it to provide new information" (Oxford Learner's Dictionary, n.d.). Predictive analytics merged naturally with large data sets. Predictive analytics uses information to determine a particular trend and predict the following action. That action may be from an individual, a group of people, weather trends, health trends, and many more. The process of geocoding in big data uses descriptive locational data as geographic references for mapping. Geocoding assigns a geographic code to link data and analysis, taking predictive analytics beyond information correlation and into the mapping of locations in relationship to human behavior. In geocoding projects, those without information are either positioned as impotent or politically aggravating (Goldberg et al., 2007). Algorithms are a set of steps to accomplish a task. From algorithms, we jump to artificial intelligence (AI).

The history of predictive analytics reveals that race, gender, and sexuality underlie the development of information technologies and fuel the anxieties toward building machine learning that can normalize futurity. Predictive analytics emerged out of World War II and the Cold War, at a time when the US armed forces were the driving force behind digital computer development from the forties to the early sixties (Edwards, 1997, p. 43). Military financial support inspired a collaboration between research universities and the armed forces, driven by a cultural belief that systems theory and cybernetics could create a utopia, a world in which populations abroad were controllable through technology (p. 73).

Military discourse positioned the futurity of America as reliant on the use of technology to control information as data. This post–Cold War era of the closed world imagined subjects abroad as uncontrollable without these technologies, representing the chaos that drove the US military to invest in further information technologies. The closed world responded to the US military and political leaders' anxieties from the overseas chaotic and dangerous spaces, which are "controllable by the powers of rationality and technology" (Edwards, 1997, p. 72).

Anxiety among the armed forces also arose around the futuristic demilitarized soldier as these technologies rapidly evolved. While some in the military were skeptical of the emerging "electronic despotism" (Edwards, 1997, p. 72) of the sixties, the Vietnam War inspired General Westmoreland to foresee an automated army, one that might elude war loss through data control (p. 72).

Computational predictive analytics could lead to US control over uncontrollable enemies abroad in a post–Cold War era but also had the potential to focus on civil rights movements through the fifties and sixties. Computing scholar Tara McPherson (2012) argues that these systems' programmers were not independent of the racial anxieties of the sixties and seventies. For white male programmers developing systems such as UNIX, the incorporation of simplicity, security, efficiency, and control paralleled the national attempt to control race and social movements via standardization. The post–World War II computational culture was not unaware of organized movements for racial justice, but computation and coding benefited security culture used for surveillance and control of communities of color domestically and abroad (McPherson, 2012).

The dot-com bubble of the nineties brought exponential growth of digital telecommunications, as computing bits of information began to drive the economy (Negroponte, 1995). In addition, natural language processing increased the capacity to organize and analyze the vast amounts of text from the internet and other sources (Lazer et al., 2009). This new economy of information led to the current state of the computational social sciences, which use data patterns to predict individual and group behavior.

The Threat of No Information

The rise of big data and the attempts to manage massive amounts of text have been framed mainly by these same narratives of the past: technology as a tool that will only improve the quality of life for specific users and help label and control those without information (Lazer et al., 2009). Technology enthusiasts advocated for big data because of its ability to organize massive amounts of information to determine human behavior. Demonstrating the hype and mass investment in big data in 2008 was Chris Anderson, editor of *Wired* magazine, excitedly mulling through the potential of big data as a source to measure human behavior on the largest scale, noting that the correlation of data can discover new species and advance biology without hypotheses. The turn toward the computational social sciences inspired Anderson to name the oncoming era of big data the Petabyte Age, where the cloud stores information and massive amounts of data, and applied mathematics is used to create tools to make conclusions through correlation: "Who knows why people do what they do? The point is they do it, and we can track and measure it with unprecedented fidelity. With enough data, the numbers speak for themselves" (Anderson, 2008). He proclaimed that big data is the end of theory because correlation now supersedes causation, "It's time to ask: What can science learn from Google?" (Anderson, 2008). At the time, that was a radical sentiment. However, it was also a forewarning of systems to come. Those algorithms would become the determinant of life chances.

Big data and predictive analytics emerge as surveillance projects, commonly justified through an argument that promises state futurity via specters, such as terrorism, that at once threaten the nation and bring it back together (Esposito, 2010). Big data names these noninformation threats as impulsive, erratic, and inconsistent with normativity. In 2007 Columbia University sociologist and principal researcher at Microsoft Duncan Watts hypothesized big data's potential in predicting potential social and economic phenomena (Watts & Perretti, 2007). He eagerly advocated that the future looked promising through social network analysis as a revolution of the internet and social sciences. Big data gathers information to quantify populations and predict futurity. Watts and Perretti (2007) dreamed up the dataless social phenomenon that can be data mined, mastered, and controlled: "From the apparent wave of religious fundamentalism sweeping the Islamic world (and parts of the Western world) to collective economic security, global warming, and the great epidemics of our times, powerful yet mysterious social forces come into play."

According to Watts and Perretti (2007), the future of that current state was uncertain, threatened by global warming and terrorism, and situated at an impasse of crisis. The mystery of the future acts as a threat to the state; the act of prediction, the knowing, is the critical first step to normativity. Academics and programmers also saw the potential for big data to assist in improving predictive analytics: "Recent literature has suggested that computational analysis of large text archives can yield novel insights to the functioning of society, including predicting future economic events" (Leetaru, 2011). The enthusiasm around big data in the early aughts was palpable.

Net Sinks and Correlated Data

Predictive analytics uses aggravated information—metadata such as personal addresses, news sources, and social media—to analyze trends that might indicate correlations in the past activity that may predict future actions. Early big data developers saw the potential of networking data around those types of people or sources without data; multiple approaches mapped subjects with little or no information.

Many factors may contribute to an individual and whole communities without information. Some may intentionally resist technological identifiers. Others may not access the necessary resources, such as residents in rural areas without typical addresses. Some may be unmappable based on their status as people without documentation (Wyatt, 2003). I use the term *correlative normativity* to refer to when AI correlates subjects with little or no mappable information (such as undocumented people without social security numbers, rural homes without a street address) with data points associated with documented information (such as nearby rural homes with a street address). Correlative normativity names gaps in data, such as undocumented people, and converts them into data-producing subjects. Subjects named with correlative normativity are not citizens with full rights but those made normative by way of correlated data. Crawford notes, "machine learning presents us with a regime of normative reasoning that, when in ascendant, takes shape as a powerful governing rationality" (2021, p. 17).

During the evolution of big data in higher education, data science scholars labeled populations without data as both unproductive and aggressive (in their unpredictability), implying excess in resisting the normative structures that predictive analytics uses to map those subjects. The details of algorithms from companies like Palantir are kept private, but we can look at early writings on surveillance systems. For example, during the Arab Spring in the Middle East, big data and predictive analytics demonstrated predicting, policing, and preventing uprisings (Leetaru, 2010, 2011).

Through data science, populations that are text-mined are called passive when enough information reveals conclusive results and leads to control (Leetaru, 2010). For example, scholars in data science have labeled countries and continents without enough networked information as unproductive: "Africa as a whole is a *net sink*, with many more reports produced about that continent than are sourced from it" (Leetaru, 2010, italics mine). The qualifier of "predictable" correlated normativity is to provide enough networked information to contribute to text mining. Affect in the Middle East and Africa, such as "public mood," is a minable data point for predicting political instability and uprising (Cioffi-Revilla & Rouleau, 2010; Goldberg et al., 2007).

In geocoding, subjects without information disrupt the mapping of geographic references, and without the information, they are therefore unpredictable. Geographic information system scholars refer to areas that either frequently change because of geographic features that are not addressable or lack consistent addressing as "hindering" data sets: "You can't grow a network with no information, so somehow there is information coming into play that's driving the shape around this into play. The research problem is always, 'what's that information that's driving this away from random?'" (Cioffi-Revilla & Rouleau, 2010). This data science scholarship calls spaces and subjects *with* mappable information passive and good because they can be predicted and controlled. But they call subjects without information erratic, unpredictable, and nonnormative. Yet subjects without information can be fixed by correlating surrounding information, enclosing them into the normative fold through their information neighbors. For example, a rural home without an address can be labeled and mined through the correlation of the data points, or home addresses, surrounding it.

With the rise of big data, software engineers and surveillance projects correlated information around those *without* information to give those identities a data point. They became mappable, documentable subjects.

Ankle Monitors

Ankle monitors are an example of how immigrant data is a tool of constant surveillance, monitoring, and fear, and how *correlated* data has become an information weapon to conduct more raids, processing more immigrants through detention centers.

Over the past fifteen years, ankle monitors and the electronic monitoring system have gained momentum and had a particular upsurge with monitoring immigrants. According to Pew Charitable Trusts, the use of electronic monitors doubled between 2005 and 2015, justified as a more humane way of reducing incarceration numbers (Sanders & Kilgore, 2018). Electronic ankle monitoring costs the person wearing that bracelet five to twenty-five dollars a day (Sanders & Kilgore, 2018). However, there are numerous flaws with the ankle monitoring system, including the monitor losing signal or connection; some convictions requiring a person ti wear the monitor for life; the inability for that person to have an MRI, mammogram, X-Ray, or CT; and the unknown and opaque process of data storage and retrieval, most likely owned by the local department of corrections or sheriff's office (Sanders & Kilgore, 2018). The ankle monitor is a visible social stigma: Wearers are judged early and often for wearing an ankle monitor and often are denied jobs because of it. However, beyond the visible lies the invisible, the high value of that monitor's data among immigrants.

Ankle monitors on immigrants began in 2002 and have increased since 2014 (Grinspan, 2019). For immigrants coming from detention centers, GPS monitoring can be a condition of their release. An NBC report on ankle monitors found that as of August 2019, there were 99,349 people in ICE's Alternatives to Detention program, and 43,233 of those were subject to the GPS tracking devices (Silva, 2019). ICE has put ankle monitors on detained immigrants, and that allowed those people to live independently. However, like much data surveillance, their data was used to raid workplaces such as A&B, Koch Foods, Peco Foods, PH Food, and Pearl River Foods (Silva, 2019). For example: In 2016, ICE released a Guatemalan from detention with an ankle bracelet. She went to live and work in Mississippi. ICE watched her data, which showed that she worked at PECO Foods' Canton processing plant for ten hours a day for multiple days a week. With her GPS data, ICE determined that they would raid PECO Foods. A similar raid happened in 2018 at Koch Foods. Immigrants were detained, given ankle bracelets, released, and then deported with a larger group of immigrants based on their networked information. A growing trend and an alternative to the ankle bracelet is a new smartphone app called SmartLINK, which uses facial recognition software to confirm identity and location tracking (Grinspan, 2019).

Ankle monitors are an example of correlated data in its earliest form. A person without a social security number, a smartphone or a social media account can be *correlated* to a networked person into these surveillance technologies.

THE LATINX IMMIGRANT DATA BODY

Current Models

For decades, immigrants' information has been classified and organized for tracking, surveillance, and deportation purposes. With new digital technologies, undocumented people could now be *known* through their information (Chaar-López, 2019). That process of data gathering is not new, but what is new is the stakes and value. Immigrant data is now up for sale to the most technologically advanced bidder. AI technologies are assumed to make predictions through large data sets that lead to epistemic truth. Alex Campolo and Kate Crawford call this *enchanted determinism:* "AI systems are seen as enchanted, beyond the known world, yet deterministic in that they discover patterns that can be applied with predictive certainty to everyday life" (Crawford, 2021, p. 213). Those unseen AI systems discover patterns to predict, locate, and document undocumented immigrants.

With the funding of the smart wall, Silicon Valley companies were urged to use their already developed technology and innovations for tracking immigrants along the borderlands. It is probably impossible to list all companies bidding for smart wall funds and tech adaptation. However, we will look at how immigrant data is proposed as a digital product in the same way that consumers' data is valued in algorithms for marketing. Your clicks, likes, eye tracking, fingerprint scans, and facial recognition scans are valuable. It is *data,* and when it is gathered into information and correlated to your personal relationship data, it can tell a company about your next purchase, what type of product holds your attention, and what type of app you are most likely to purchase.

The smart wall technology is similar. Technology is designed to track a person with little data. Like the ankle monitors, they may correlate that person according to someone *with* data (more about that in chapter 5), but more than likely, they will need to find that person in the borderlands with little to no data. That person's *body* is the data; it is the goal around which the technology is designed.

AI and Latinx Immigrant Silhouettes

Media coverage of the development of surveillance technologies rarely names Latinx immigrants as the direct object of desire. Instead, the Latinx immigrant in technological design is a specter. Fear and anxiety around Latinx immigrants are always present, but *naming* them in the pitching, funding, and promoting of technological design is cloaked.

In 2018, with Palmer Luckey's break from Facebook and Oculus, he turned his attention toward a new venture: military technology, particularly along the US-Mexico border, through various hardware and AI. Lattice is a software system that takes data from the technologies scanning for data, such as the Integrated Fixed Towers, satellites, and cameras along the border. Then it uses data mining to correlate that information and interpret it for border patrol. The difference between Lattice and previous technologies on the border is that it can distinguish between noises in the borderlands such as animals and vehicles or humans on foot (Dean, 2019). Once Lattice determines if data is a source to investigate, the user can use VR to investigate that geographical location on the border.

As described by Anduril: "Lattice uses technologies like sensor fusion, computer vision, edge computing, and machine learning and artificial intelligence to detect, track, and classify every object of interest in an operator's vicinity" (Anduril, n.d.). AI is discursively positioned as the missing key in a prolonged attempt to locate immigrants in the borderlands. Hardware devices such as ground sensors, drones, towers, and cameras feed in data, and algorithms work to determine if an object or movement is a person of interest, in turn alerting border patrol to that location. In addition, AI is sold as overcoming the previous problems of false triggers, discerning between an animal, an agent, or an immigrant. In these models, Latinx immigrant bodies are the information product to be captured.

Coolfire Core (previously named Coolfire Solutions) is one of many companies proposing AI models for the digital wall. According to Coolfire, "We help people operating on the frontlines, struggling to make decisions, challenged with doing what is right. We empower everyday people to be everyday heroes" (Coolfire Solutions, n.d.). Coolfire describes the Digital Wall as "a row (or rows) of in-ground sensors—cameras, infrared, and motion detectors—which rather than actively preventing illicit border traffic, alert and enable human security teams to take action" (Coolfire Solutions, 2018). In this model, Latinx immigrants are the unpredictable object with no data, visualized as no more than a stick figure. Coolfire Core proposes a Situational Awareness Platform to make the undocumented subject a capturable information object.

Coolfire's Situational Awareness Platform is an approach that brings together data from an environment (named Perception), turning it into information (named Comprehension), and converting it into a prediction on what will happen next (named Projection). This platform builds off what already exists in predictive analytics, long used for military models: " . . . situational awareness involves picking up cues from the environment, putting those cues together to understand what is going on, and using that understanding to predict what may happen next" (Coolfire Solutions, n.d.). Figure 1 reproduces this model as a visual aid. This platform demonstrates how to make unknown objects into predictable subjects.

In the borderlands, there is enough infrastructure to network that data together and, with the help of AI, bring the Latinx immigrant from an unnamed silhouette into a more precise subject. AI *builds* Latinx immigrants in borderlands from correlated data, from unidentified objects, into a more precise data body, a monetarily valuable information product.

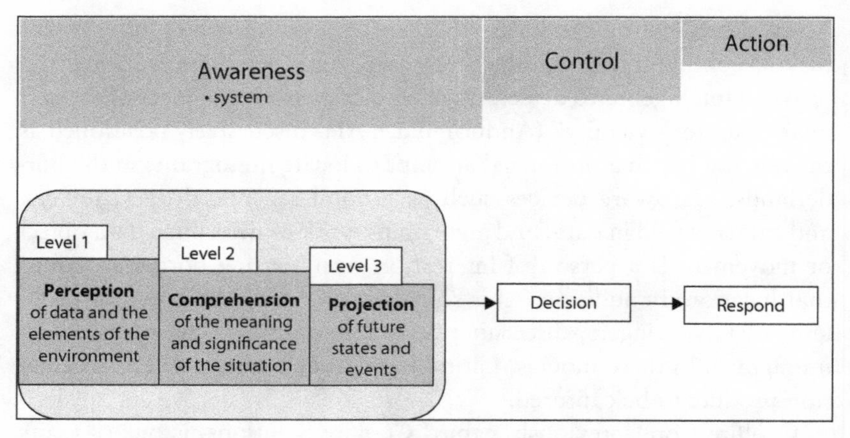

FIGURE 1. Endsley's situational awareness model as described by Coolfire Core. Illustration by the author.

An NBC News interview with cofounder Brian Schimpf of Anduril demonstrates the function of new technologies in the borderlands of Southern California. The NBC correspondent starts the segment with an image of a man walking through familiar borderlands: dry brown shrubs under his feet, mountains eclipsed by the noon California sun. Sensors and cameras identify his location, and a drone called Ghost flies in closer to the man, seemingly dropped in the middle of an isolated desert. However, the report cuts away from the Latinx man crossing the border and into a discussion of how these new drone models can knock commercial drones out of the sky, essentially taking down any military or domestic threats (Pettitt et al., 2018). The narrative structure here is clear: in exchange for heavy borderland surveillance and centralizing Latinx immigrants as the object to capture, emerging tech companies will bring citizens safety through technology. We will see that this message repeats regarding developing technologies that center immigrant surveillance—tech in exchange for an undeniable social good that is seemingly impossible to oppose. Luckey validates this work by calling on a familiar metaphor: "Right now the skies are the wild west. Anyone can do what they want as long as they're willing to violate the law, and no one can stop them" (Pettitt et al., 2018). The altruistic uses of technology services to justify casting the all-seeing gaze on Latinx immigrants in the borderlands, one of the many rogue groups invoked in the metaphor of the Wild West. The design centralizes Latinx immigrants as the object to be found and converted into identifiable information for these technologies. Just as that drone is programmed to find other drone threats, it is also programmed to find Latinx immigrants.

These models—the drones, ground sensors, VR, AR, situational awareness platforms, and AI—piggyback off the original goals of the early sign cutting along the border. They aim to make technology sophisticated enough to track and catch Latinxs. In the early days of sign cutting along the US-Mexico border, tracking was an avenue for INS and border patrol to touch base with their instincts. In 1934 an INS deputy commissioner wrote: "There is one angle to the work of the Border Patrol which links it with the Indian fighters of the early days. Inspectors use the science of tracking. Broken reeds on a riverbank may tell the plain story of the landing of a smuggler's boat'" (quoted in Stern, 2005, p. 76). The past reflected future concerns: The Southwest borderlands' topography and the environment were a puzzle that must be solved. In the thirties, border patrol attacked that problem to develop what they considered a "sixth sense" (p. 76). However, in the twenty-first century, the aim is precise. Ultimately, those technological infrastructures work together. Through AI, patrol officers learn the same signifiers that once taught them how to distinguish a human from an animal in the desert; or more specifically, how to distinguish the Latinx immigrant from a citizen or patrol officer. That programming, like sign cutting, racially profiles in that the purpose is to distinguish between a Latinx immigrant in borderlands and all other targets.

Machine Learning Feedback Loop: Consumer-Immigrant-Consumer

Latinx immigrants as data are valued for the moment they are captured on the border, and that data can go on to improve other systems related to that surveillance net. For example, the more products I look at through Amazon or Google, the more algorithms are refined to tailor future ads and product placement. Our data is part of improving those algorithmic systems in the *machine learning feedback loop* (Singh, 2019) (fig. 2).

Critical computing scholar Sava Singh gives the example of Gmail accounts that refine autocorrect and predictive writing within an email. When Gmail suggests autoresponses of "Great!" or "Sounds good," those suggested responses change us, and we may change our responses or search results based on that algorithmic feedback. Immigrant data feeds into multiple tracking systems on the border; those systems learn

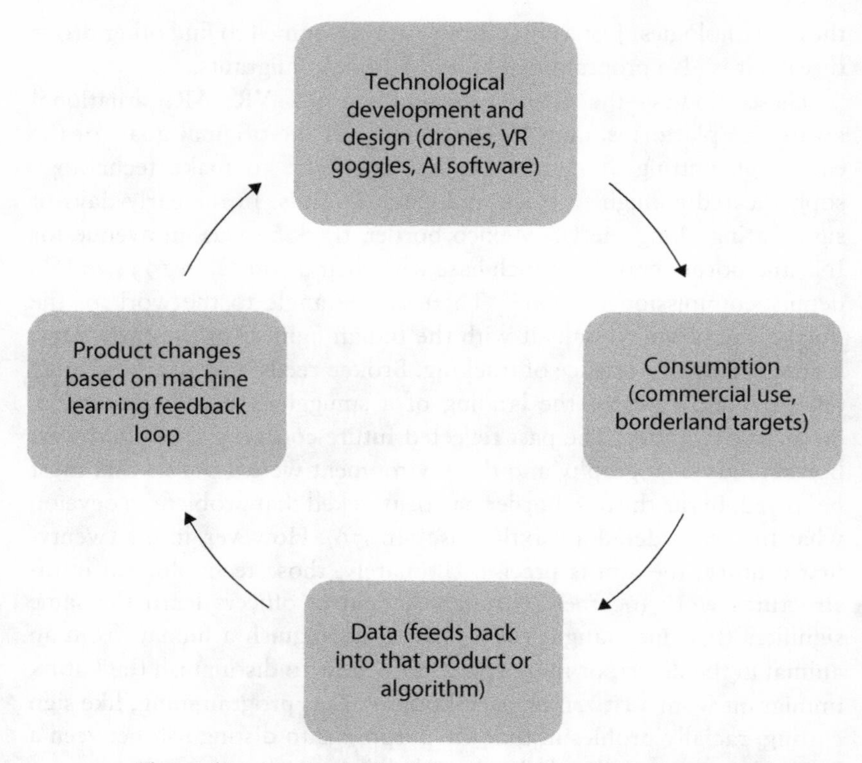

FIGURE 2. The machine learning feedback loop as applied to borderland technology. Illustration by the author.

from that data, refine themselves, change the product based on it, and return it to the commercial market, as demonstrated in Figure 2.

This machine learning feedback loop is evident in Anduril Industries' mission statement:

> . . . consumer and commercial technology has outpaced the defense industry in both capability and cost. In addition, developments in AI, drones, and sensor fusion have thrived in the private sector resulting in the commoditization of the same advancements that will win and deter future conflicts.
>
> We are a team of experts from Oculus, Palantir, General Atomics, SpaceX, Tesla, and Google exploiting breakthroughs in consumer and commercial technology to evolve our defense capabilities radically. (Anduril, n.d.)

Mission-based language of merging Silicon Valley with military defense is strengthened by embedding immigrant data and citizen-consumer data. The machine learning feedback loop is what Luckey has created to compete with China's military tech development.

FIGURE 3. A landscape of trees and terrain rendered with LiDAR imaging. Image by black_mts, via Adobe Stock Photos.

Light detection and ranging (LiDAR) technology is an example of the machine learning feedback loop. The company Quanergy Electric, using 3-D technology built for self-driving cars, proposed to build a virtual border wall from LiDAR sensors for surveillance systems, which can detect movement in a 360-degree radius and two hundred meters of the sensors. In a CNBC report on the virtual border wall, Quanergy demonstrated this technology, revealing how LiDAR technology "classifies objects and can distinguish between a human, vehicle or animal" (Bundy, 2019). The graphic visualizes people in the traditional animated stick figure form, as white, cartoonlike stick figures moving across a dry Texas desert (CNBC, 2018). Here we see the shadow of the Latinx immigrant again. Rarely are they directly named by these tech firms beyond the design of technology, such as LiDAR: To bring the Latinx immigrant in borderlands from cartoon outline into complete data form, and to be able to discern among animals, other agents, and immigrants.

Tracking Latinx immigrants becomes one of the products LiDAR generates. Lidar, drones, VR goggles, and AI along the border become part of that machine learning feedback loop. Those technologies are then improved, and that improvement enhances the commercial use of such tech as the self-driving car. While Latinx immigrants are rarely named beyond drug dealers, cartel members, and the Central American caravans that move toward the borders, their data becomes the main information product. That data will go back into improving those technological products for a consumer market—embedding the Latinx immigrant into the technological design. Figure 3 demonstrates how images appear through LiDAR technology.

In these models, undocumented people are cast as a threat to citizens for their inability to be mapped, traced, or provide information in the acquiescing way of citizen-consumers. The promise of these surveillance technologies and data gathering recasts undocumented people as nonthreatening if they can be predicted and documented as data. However, there are even more alarming emerging industry models that employ immigrant biological data as a catalyst for technological development.

Biometrics: The Next Next Frontier

Companies such as BI2 Technologies, Amazon, and DNA testing companies have a financial stake in gathering the biological data of undocumented people at ports of entry and borderlands. Biometric technologies and biological data have increased in value with the sophisticated development of scanning and storing the body. In 2019, the Office of Biometric Identity Management (OBIM) estimated that DHS will have conducted face, fingerprint, and iris scans of at least 260 million unique identities in the biometric database by 2022 (Rohrlick, 2019). While the dehumanizing conditions that undocumented people face in US working conditions and detention centers are not new, the growing collection of their biometric data is a new development.

What is biometric data? According to the Biometrics Research Group at Michigan State University, biometric data is "the automated recognition of individuals based on their anatomical and behavioral characteristics such as fingerprint, face, iris, and voice" (Jain et al., 2011). Biometrics identifies people's physical and behavioral characteristics and includes palm prints, gait, and tattoos. Biometric data has long been used for forensic testing and is collected during processing people into jails and prisons, leading to large databases of fingerprint and iris scans. The collection of diverse biometric data—through fingerprinting and iris scans—is valuable to the authorities because it yields more accurate results with more nuanced data. For instance, a fingerprint can yield mismatching results, while an iris scan can be more precise.

Biometric Data

Fingerprints, as identifiers of an individual, have long been used to represent the unique signature of a person; biometric history is anchored in colonization and incarceration. As early as 500 BC, Babylonians would use their fingerprints to conduct business with clay tablets (Mayhew, 2018). In 1858 Sir William Herschel, a part of the colonizing military in the Civil Service of India, used fingerprinting to document his employees.

In mid-nineteenth century, France systematized fingerprints with "'anthropometries," a field of study that measured the physical characteristics of criminal offenders and their punishment (Mayhew, 2018); criminal anthropometry is the foundation for biotechnologies. In 1903 the New York State Prison began using fingerprinting for "the identification of criminals. . . . In 1904 the fingerprint system accelerated when the United States Penitentiary at Leavenworth, Kansas, and the St. Louis, Missouri Police Department both established fingerprint bureaus" (Mayhew, 2018) In the nineteenth century, through scientific procedures believed to be authoritative, body shape measurements could predict and reveal who would be a criminal. Anthropometry and fingerprinting were eugenic tools that argued that the inferiority of some races and the features of criminals were directly linked to their measurable physical attributes. Fingerprinting and facial features were distinct tools of anthropometry to identify criminals in the "bertillonage system," a system that "used identification cards with detailed descriptions of a criminal's bodily features" (Wevers, 2018, p. 98).

In 1936, ophthalmologist Frank Burch hypothesized the use of iris patterns for the unique recognition of individuals. In the sixties, mathematician and computer scientist Woodrow W. Bledsoe created facial recognition for the US government to locate eyes, ears, nose, and mouths in photographs (Mayhew, 2018). In 1969 the FBI contracted with the National Institute of Standards and Technology to automate fingerprint identification, and by 1975 the FBI had developed scanners and new technology to read fingerprints easily. In 1988 the Los Angeles County Sheriff's Department began using the first semiautomated facial recognition system for digitizing mugshots of both photos and composite drawings. In 1994 Lockheed Martin was selected to build the FBI's Automated Fingerprint Identification Systems. The nineties were filled with biometric breakthroughs and applications, from hand geometry recognition at the 1996 Atlanta Olympic Games to the use of a fingerprinting system at airports to bypass the lines (Mayhew, 2018). Finally, early in the next decade, facial recognition became the center of innovation—used en masse at the 2001 Super Bowl in Tampa, Florida, for scanning individuals in the stadium.

The Built-in Bias of Biometric Data

Advocates often consider biometric data infrastructure to be a flawless system compared to informational identity. For example, a social security number or name can be falsified, but according to biometric data advocates, a facial feature, fingerprint, or iris scan cannot be fake. Since those technologies are naturalized as value-free, biometric data is assumed to be an absolute identifier: "biometrics equate the body with identity—and the body thus becomes the most important signifier of identity" (Wevers, 2018, p. 91). But biometric identity neglects the social experience of identity—subjects' experiences of the world, of power, and intersectional stratifications are not considered in biometric identity (p. 91).

Built-in bias to biometric data was revealed early on by researchers Buolamwini and Gebru. In 2018, their MIT research group found that three facial recognition systems were consistently misclassifying darker-skinned women, with an error rate of up to 34.7 percent. In contrast, lighter-skinned males had an error rate of 0.8 percent (Buolamwini and Gebru, 2018). With large databases of biometric data, that type of error leads to further racial profiling by way of AI. However, with the justification of technologies as neutral and progressive, those systems justify further racial profiling as a means of progress. Researchers also found that the systems were racially biased/racist because the programming used larger data sets of light-skinned males. The programming itself was biased/racist, and therefore programmed in racism. Biometric programming and machinery are built around whiteness as the base for facial features and skin tone. As a result, biometrics often fail "at the intersection of racialized, queered, gendered, classed, and disabled bodies" (Magnet, 2011, p. 47). For example, biometrics has failed to read Asian women's fingerprints and the calloused fingerprints of laborers (Wevers, 2018, p. 92). These programmed biases and failures lead to racial profiling in highly visible and surveilled spaces (Wevers, 2018), and they demonstrate the erosion of biometrics as a fail-safe way of identifying individuals.

Biometrics in the Borderlands

In 2017 the ACLU reported on the adoption of iris technology along the US-Mexico border. Thirty-one sheriffs there had begun adopting iris-scanning technologies developed by BI2 Technologies. Iris scanning has long been hailed as a more accurate identifier compared to fingerprints. BI2 also stores that biometric data (Cagle, 2017). However, the iris scan has been prone to critique because it encourages racial profiling along the borderlands. Brown eyes are the target for scanning, so those people in borderlands with brown skin and dark hair are justified by law enforcement as the target for that biometric scanning. The BI2 iris database is open to 180 law enforcement clients of the Southwestern Border Sheriffs' Coalition ("Southwestern Border Sheriff's Coalition (SBSC) to immediately begin improving," 2017).

As is typical with these developing technologies, they apply to all people apprehended on the border, but the Latinx criminal justifies them. Sheriff Joe Frank Martinez of Texas described the audience for this tech: "Every day, the Sheriffs and their professional staffs confront drug smuggling, human trafficking, stolen vehicles and firearms, crimes against persons, crimes against property and violent crimes such as murder and sexual assaults in their countries" ("Southwestern Border Sheriffs' Coalition," 2017). Because of the fear of violent criminals, iris scans became a valuable database.

Biometric identification has increasingly become part of the key to accessing human rights for refugees around the world. Again and again, authorities have turned to DNA testing due to a crisis. War, violence, cartels, and starvation motivate people to leave their homes and everything they have known to find a haven for themselves and their families. For over a decade, the US State Department has used DNA to test familial relations among refugees for family reunification programs. However, within the history of biometric testing of immigrants, we see that their crisis status is a gateway into further developing technologies and surveillance that document that group. DNA becomes the very last frontier of documenting the minutiae of the immigrant data body. It is a way of determining who a person is through science. Furthermore, advanced rapid DNA testing techniques have become more widely available and apply to all immigrants. DNA testing is a bottom-line biometric avenue, proclaiming: this is who you are; this is the group you belong to.

The State Department began DNA testing on immigrants in 2008 in Nairobi, Kenya, to determine false relationships among family members. The Refugee Family Reunification (Priority Three) P-3 program allowed individual cases from eligible nationalities access to the United States for family reunification. The State Department claimed that most immigrants from Kenya, Ethiopia, Uganda, Ghana, Guinea, Gambia, and Côte d'Ivoire (US Department of State, 2008) were falsely claiming family members for citizenship. However, this analysis did not account for nonheteronormative familial relationships, adopted children, and chosen family members. As a result, DNA testing quickly became a fixed way of determining whom refugees were legally allowed to reunite with; DNA is the emerging form of immigration documentation. Since 2015, for Syrian refugees to gain access to the United States, they have had to demonstrate familial relations through DNA testing. This has shaped a heteronormative ideal of the immigrant family: those related through DNA were given access to each other, but adopted children and family members were denied access as refugees (Worth, 2015).

DNA testing is used as a blueprint to grant access to some people in the United States, and biometrics are increasingly used as the key for passing into safe havens. In 2013, iris scans were increasingly used on Syrian refugees as part of their application process. The United Nations built a sizeable biometric database of 1.6 million refugees' eyeball scans (Worth, 2015). During this time, rapid DNA testing was developed. Rapid DNA testing promised results in ninety minutes with a Q-tip swab of the mouth. The Rapid DNA testing of this time was used solely to confirm kinship between two individuals (Worth, 2015).

In 2019 Homeland Security began collecting biometric information on every refugee who applied for resettlement. In January 2019, the United Nations High Commissioner for Refugees (UNHCR) collaborated with USCIS to share fingerprints, iris scans, face images, and other biometric data. They stored it in Homeland Security's Automated Biometric Identification System (IDENT), which houses more than 250 million people's biometric data. The Homeland Advanced Recognition Technology system now replaces IDENT (Corrigan, 2019). Explicit consent is not obtained to gather that data, and the information is shared with multiple agencies.

In 2018, news broke about families in immigrant detention centers; no records had been kept on reuniting those families. So the DNA testing groups 23andMe and MyHeritage offered genetic testing kits to reunite those families, guaranteeing that they would protect the privacy of those tested. Then, as though a lightbulb went on at ICE, in May 2019 the agency began using rapid DNA testing at the border to allegedly identify individuals who were not related through biological parent-child relationships.

ICE has contracted ANDE, a Massachusetts-based Rapid DNA testing company, to manage the data for those kits. ANDE is a firm founded out of MIT in 20014 in the Greater Boston area. ANDE's products are generally biotechnology and aimed at both government agencies and industry. Like many of these technologies, the justification for collecting immigrant biometric DNA is done through criminalization. On their corporate About page, ANDE states, "By reducing crime, exonerating the innocent, monitoring borders, and reuniting families, Rapid DNA represents a fundamental opportunity for the public and private sectors to protect the innocent" (ANDE, n.d.). Along with immigration, other services that ANDE represents are DNA analysis for human trafficking and disaster victim identification.

Consumers of genetic testing, such as 23andMe, can opt in to data research and sharing. However, privacy law does not protect consumers' full ownership of DNA information. The Genetic Information Nondiscrimination Act (GINA) of 2008, which solely protects consumers from being discriminated against based on their genetics, primarily from employers and health insurers, is the most comprehensive policy covering DNA. State laws, too, attempt to protect against genetic discrimination, such as the California Genetic Information Nondiscrimination Act (CalGINA), to extend protection in emergency medical services, housing, mortgage lending, education, and other state-funded programs. However, the DNA testing at the border conducted through federal agencies can supersede state laws and the protection of sanctuary movements.

Courts have not yet determined who owns the biological data of undocumented people. Arresting and deporting undocumented people is no longer done solely for those purposes. The apprehension of the undocumented person is now lucrative because their biological, biometric data can be mined, harvested, and used to generate profit.

The use of DNA analysis is justified as the most accurate way of determining familiar relationships. DNA analysis and results are used as a map to determine who people are and where immigrants belong. It can be viewed as a key, but it is a key that blocks access to countries and privileges while solidifying an immigrant's identity.

Perfection around Ethnicity, Eugenics

Biometrics, inevitably, has a history intertwined with eugenics. Eugenics "fixes" identity by assigning fixed attributes to measured physical features; biometrics fixes identity by assuming neutral programming and assigning a measurable value to physical features. After it is used to determine who is valuable, that biometric information becomes a screening tool for what type of identity society values. Benjamin notes the move by genome organizations to study intelligence in DNA sequencing, to be able to isolate those genetics and choose to reproduce according to preferred genetic sequencing (2019, p. 114). Biometrics justifies inclusion through identity: "As it is constructed through rigorous screening regimes in geopolitical contexts today, the biometric body is intended to become precisely an undoubtedly biophysical source of certainty about an individual's identity, which will allow for the determination of his or her societal statuses" (Liljefors and Lee-Morrison, 2015, p. 55).

In many ways, it is presented as the bottom line. DNA is the genetic blueprint of a human. However, DNA becomes a key when given social value—such as ethnicity, gender, health, and familial relations. In 2018, while his administration was rolling back rights for transgender people, particularly their inclusion in the military, President Trump's memo mentioned that a genetic test would be used to determine a person's biological gender as undeniable evidence. Transgender rights activists and scientists have long argued that the genetic X and Y chromosomes are not a determinant of a person's gender identity. By way of genetic testing, science became a form of evidence to fix a particular group of people's identity and their belonging and inclusion in specific organizations, spaces, and accesses to rights. Trump was drawing on a historical tradition in the United States of "proving" stereotypes and xenophobic beliefs by ways of science, particularly genetics.

Eugenics has established roots in the United States. However, traditional eugenicists usually invoke Nazis and the eradication of ethnic groups on the basis that those groups are genetically inferior. In the United States, eugenics has been more naturalized through sterilization, immigration laws toward preferential ethnicities, and strict border enforcement on the US-Mexico border to restrict Mexican influx into the population.

The US eugenic practices against Mexicans at the border began with the Texas-Mexican border and a virus outbreak. In 1916, in order for Mexicans to cross at the Laredo–Nuevo Laredo entry point, they were bathed and physically examined by the US Public Health Service (PHS). Once met with approval, they got stamped with the word "ADMITTED" on their arms (Stern, 2005, p. 57). At that time, Laredo PHS medical inspector H. J. Hamilton defended this practice as defending the United States against "lice, smallpox, and other germs" associated with Mexicans through stereotypes (p. 57).

Eugenics policies toward Mexicans were preemptive, before they crossed into the United States, through medicalization and militarization. In 1915 a typhoid outbreak in Mexico led PHS to prepare the border with quarantine disinfection plants. In 1917 the agency announced that anyone from Mexico must submit to kerosene showers and have their luggage sulfur-fumigated. Beyond typhus, any Mexican woman with head lice was subject to a kerosene and vinegar treatment, and men were subject to having their hair cut off completely. After being bathed, vaccinated, and having their clothes and baggage disinfected, Mexicans had to pass through psychological profiles, questioning their visas and citizenship. They were then given a certificate of medical examination that read "United States Public Health Service, Mexican Border Quarantine" (Stern, 2005, p. 62). Anti-disease delousing and fumigation continued from 1917 to before World War II. In contrast, those crossing the Canadian border and Ellis Island were not subject to such racializing conditions. "To a great extent, the pathologizing of Mexicans represented an extension of the association of immigrants with the disease into new racial and metaphorical terrain" (p. 67).

Eugenics advocates would go on to influence immigration policy to cap Mexican immigration, all with an eye on the potential for Mexicans to corrupt the semblance of a white, perfect American family. In the early twentieth century in California, eugenics movements were gaining momentum against Mexicans and African Americans in particular. As early as the thirties, political rhetoric railed against Mexican family size and use of public assistance, and hospitals were sterilizing Mexicans at alarmingly high rates (Stern, 2005).

Biometrics are used for data tracking, surveillance, citizenship qualifications, borderland tracking, deportations, and racial profiling; gathered through fingerprints, face scans, tattoos, iris scans, and DNA. The uptick in DNA gathering is defended to reunite families and identify false claims to families. These uses of biometrics are a way of gathering large data sets, for example, many brown eye iris scans, and storing them in a database for recognition purposes, therefore leading to further policing and deportation. Biometrics fixes identity by way of measuring features and gathering them; eugenics takes that fixed identity and justifies the removal of groups of people based on their physical features and unsuitability for citizenship. The use of DNA at detention centers and ports of entry is a way of saying that a particular group of people does not belong in this country based on their genetic code. It is justified through family relations, but ultimately it is a bottom-line biometric that demonstrates that those immigrants are not members of the US citizen family writ large.

Simone Brown (2015) notes that branding in the slave trade was a way to make the body legible as property, an early "social sorting" of classifications. Brown calls modern biometric racial classifications, such as these fingerprints and retina scans, "digital epidermalism": "as what happens when certain bodies are rendered as digitized code, or at least when attempts are made to render some bodies as digitized code" (p. 109). We see something similar in the borderland technology timeline. Sign cutting or classifying Mexican immigrants' gait to track them is digitized through many information systems and processed as data that can be mined and used to make inferences for future predictions.

The detailed classification of immigrant biodata demonstrates how information tech companies, in partnership with DHS, ICE, and CBP, uphold these data systems as "truths" that can separate citizen belonging and unbelonging. On classification schemes, Crawford notes: " . . . artifacts in the world are turned into data through extraction, measurement, labeling, and ordering, and this becomes—intentionally or otherwise—a slippery ground truth for technical systems trained on that data . . ." (2021, p. 127). Classifying the most granular of immigrant data "justifies" these systems as upholding the ultimate truths about that subject being: biodata as a system of classification leads with claims that it can sort the citizen from the immigrant.

This next chapter will look at the networks that gather and track immigrant data and how they are anchored to the more extensive United States system of incarceration and surveillance. This network is essential in understanding the larger picture of how information systems value immigrant data and function as the new Migra.

CHAPTER 3

Networked

Meet the New Migra

I remember being in the play area of the apartments where we lived, and I would see a lot of men running. One time my mom grabbed us when they were running around the apartments. They [border patrol] were asking us questions. That was every week, they would park outside the apartment complex. It was Sundays or during the week when most men came from work. I would see them in the streets, close to the schools when they were going to pick up their kids.

They stop people based on the color of their skin, how they dressed. They would target the guys dressed like cowboys or dressed in construction boots.

—Juana

One string of technology and surveillance leads geographically across the country and globe, and it moves back in time. This data border is not a stand-alone phenomenon but a network dependent upon older physical infrastructures, evolving because of networked tech companies, technologies, and government agencies that work together. Furthermore, for all residents in the United States, detention and deportation happen due to data correlation, or data working in the community with other data.

The players in this network are vast monoliths in the US government and the technology industries. ICE awards contracts to significant information technology (IT) companies; however, ICE and DHS are not the only agencies involved. These companies are building off the expansive so-called criminal databases that US law enforcement at the local, state, and federal levels has been using for decades, some of those managed by

Palantir Technologies, a company known for targeting and policing communities of color.

Networks strengthen emerging data borders. We have seen networks come together in the infrastructure along the physical borderlands, with that surveillance tech building upon old systems and looping in new systems to identify immigrants. Networks collaborate to turn the undocumented Latinx immigrant into a documentable, traceable data body. Moreover, networks track undocumented people's data in databases throughout the country.

As technology and data have become central to our economy and way of life, the network concept is reshaping our society. In the seventies, Daniel Bell theorized that knowledge and information were becoming one with the creation of large-scale networks. Bell (1979) foresaw that large data banks would interconnect all information and that a user could rapidly retrieve that information. Manuel Castells (1997) called the society reliant on networks and the physical infrastructure of information technology a "network society." We are neck-deep into this network society that scholars were observing decades ago. Information, formerly constrained by space and times, is now accessible and retrievable within seconds. As information systems—technological advancements, the internet and access to it, data storage, and retrieval—converged, this network society emerged and became a powerful force in our everyday lives. The information network became more central to our everyday lives than physical spaces (Castells calls this the "space of flows"), and time became shaped by and around technology (Castells, 1997).

Silicon Valley tech firms such as Palantir, Amazon, and Anduril step into a highly networked information society where information is the currency. Their ability to link databases for ICE's surveillance network is where their strength and uniqueness lie. According to the report *Who's behind ICE?*: "this interoperability has effectively expanded the reach of immigration enforcement by rendering detentions and deportations more likely to occur" (National Immigration Project et al., 2019, p. 3). This chapter gives the nuts and bolts of how these networks work together to inform CBP and ICE, and it is a request that when we see certain logos, such as that of Palantir, we visualize that white van with the green stripe. The strength in these emerging data borders is the network. ICE, federal and state systems, and commercial and military tech industries correlate those Western terrains and people with data points.

THE OLD MIGRA, THE NEW MIGRA

These networks are a new type of Migra that has expanded beyond geographic borderlands and into domestic and private spaces through data. These highly networked Silicon Valley companies are part of border patrol, La Migra. They are metaphorically deputized as the border patrol and the Texas Rangers before them. The loophole that allows private corporations' intellectual property rights to all *our* data is how companies separate themselves from La Migra. The emerging Military-Industrial Startup Complex is increasing partnerships between the military and Silicon Valley. The National Immigration Project, the Immigrant Defense Project, and Mijente name this network the "Cloud Industrial Complex" due to the many tech companies that have Federal Risk and Authorization Management Program (FedRAMP) authorizations; FedRAMP is the government program that provides security and risk assessment to cloud services. "The cloud plays a critical role in the DHS immigration enforcement system. Most key data systems supporting immigration enforcement at DHS are either hosted on commercial cloud providers or being migrated to them" (Naional Immigration Project et al., 2019, p. 5). I coin the term *Military-Industrial Startup Complex* to encompass the hardware, software, commercial products, classification databases, and cloud services that network many entities with ICE and border patrol. Migra has historically represented border patrol, ICE, and law enforcement that contribute to deportations. CBP and ICE have developed from a small organization to a vast network over the years. This new shift represents further interconnections into commercial markets by way of citizenship anxieties of immigrants at the border. A 2015 *Wired* article explicates this emerging trend and tensions around this industry in the article "How to Get Startups in on the Military-Industrial Complex." In it, author, career military officer, and Brookings Institute fellow Jason Tama gushes: "The good news is something's starting to give. The boundaries between traditional public defense goods and private goods are blurring, particularly as technologies like drones, robotics, artificial intelligence, and advanced cyber security tools penetrate further into society" (Tama, 2015).

The expansion of the United States into the West led to the establishment of the border patrol. Colonizers were expanding into the northern Mexican province of Coahuila y Tejas and expanding along the Colorado River. The Texas Rangers became a self-deputized law enforcement group that protected Anglo-American settlers in Texas in their land theft and disputes against Mexicans (Weiser-Alexander, 2020). Anglo-American colonizers eventually fought a war in 1836 against Mexico, and the United States annexed the Republic of Texas (Lytle Hernández, 2020, p. 21).

Congress established the United States Border Patrol on May 28, 1924, housed under the Immigration Bureau (US Customs and Border Protection, 2022). The Act of February 27, 1925 (43 Stat. 1049–1050; 8 USC 110) gave border patrol the power to "arrest any alien who in his [an officer's] presence or view is entering or attempting to enter the United States in violation of any law or regulation made in pursuance of law regulating the admissions of aliens, and to take such alien immediately for examination before an immigrant inspector or other official having authority to examine aliens as to their rights to admission to the United States" (US House of Representatives, 1999).

Border patrol could arrest people and search vessels without warrants. In the early 1900s, the US-Mexico border was over two thousand miles. As a result, police forces were built up inland in the United States, from Los Angeles into Arizona, all the way up to San Luis Obispo (Lytle Hernández, 2010, p. 37). In the early years of border patrol, there was considerable overlap from Immigration Services' Chinese Division, at a time when Clifford Perkins, the Mounted Chinese Inspector, was appointed to build a police force within border regions that would broadly enforce US immigration restrictions (Lytle Hernández, 2010, p. 37).

Immigration enforcement focused on Asian immigrants in the early establishment of the US-California borderlands (Lytle Hernández, 2010). In the 1930s, anti-Asian sentiment in Mexico's northern states and borderlands pushed Asian immigrants into the United States. Mexico also collaborated with border patrol to keep Mexicans from crossing the border and keep Mexican nationals within Mexico. By the late 1930s, Mexico and the United States focused heavily on policing Mexican immigration across the border. At that time, Mexico established its own form of border control to keep Mexican nationals from crossing into the United States and collaborated with the United States to implement defense. Checkpoints at train stations along the US-Mexico border and throughout the country became common, and Mexican officers enforced the provisions of US immigration law (p. 94). The 1926 Ley de Migración imposed a fine from one hundred to one thousand pesos on people attempting to cross the border (p. 96).

World War II moved border patrol from solely managing Mexican immigrants at the physical border into more broadly moving and managing various peoples around the United States. Border patrol was responsible for physically moving Japanese Americans into the internment camps run by the War Relocation Authority (Lytle Hernández, 2010). In New York, border patrol interrogated German and Italian nationals (Lytle Hernández, 2010). Border patrol beefed up its numbers through the 1940s and managed the Bracero Program, the bilateral agreement that Mexican workers would work short term on expanding US farmlands. In the fifties, as the physical border increased with law enforcement, Mexicans were dying more frequently from the harsh conditions and evading La Migra through the Rio Grande and large areas of the desert (p. 132). Border patrol engaged the natural landscape as a weapon against crossers.

Racial construction during the mid-twentieth century both included and excluded Mexicans in formations of state. Government officials labeled Mexicans as white, with access to white public facilities where they were segregated, distinguishing Mexican Americans, and undocumented Mexican immigrants through racialization. However, Mexican Americans were often subject to anti-Mexican violence in schools and public spaces. Undocumented Mexican immigrants could also use facilities that excluded African Americans. At the same time, they were subject to deportations and a border patrol torture practice of "head shaving" before they were deported (Lytle Hernández, 2010, pp. 148–150).

Operation Wetback of 1954, led by the US Border Patrol, claimed one million deportations in one summer (Lytle Hernández, 2010, p. 173). The fifties also saw border patrol appealing to the public's empathy by advocating that deportation was a human rights issue: undocumented immigrants, they advertised, were living in cramped working conditions, with inadequate food and water. "The narrative of democracy and deliverance littered Border Patrol correspondence records during the early 1950s . . . The Border Patrol's job rested somewhere between protecting the simple-minded and exploited workers from abusive farmers and protecting American society from the dangerously backward 'wetback'" p. 177). Border patrol used their "compassionate" approach toward immigrants as a means of painting farmers as exploitive landholders. Operation Wetback of 1954 was the militaristic method of liberation for all (p. 182).

In the late fifties, Immigration and Naturalization Services (INS) encouraged agents to start a campaign of criminalizing all people crossing the border, using the labels "deportable alien," "criminal alien," or "border violator" (Lytle Hernández, 2010, p. 206). They avoided deporting women and children and focused on criminalizing Mexican men (p. 206). In 1956, the INS Southwestern Region launched the Criminal, Immoral, and Narcotics (CIN) program, targeting Mexican nationals already living in the United States. The CIN program was a system of information gathering and sharing between immigration investigators and border patrol, focusing on Mexicans living in the United States. CIN paid informants for information on immigrants engaged in criminal activities on the border. While CIN publicly stated they were focusing on drugs, they privately focused on prostitution, fingerprinting, and photographing brothel owners. However, INS had little evidence that many undocumented Mexicans were criminals (p. 209), so they criminalized their arrestees and inflated the data. In addition, the agency built more detention facilities to house the so-called criminals that they were invoking in the public imagination. Meanwhile, the border patrol budget went from more than $7 million in 1954 to $11 million in 1955, to exceeding $12 million in 1956. Despite their campaign that they had ended the "era of the wetback" with Operation Wetback of 1954, border patrol and INS found that criminalization and changing their role to that of crime control of Mexicans in the United States was an avenue that could inflate their operating budgets (p. 211).

In the eighties, patrol officers were increasingly trained in emerging information technologies, including infrared night-vision scopes, seismic sensors, and modern computer processing systems (US Customs and Border Protection, n.d.). In El Paso in 1993, agents used technology in targeted, concentrated areas as a force against heavily immigrant traffic. In 1994, Operation Gatekeeper was implemented at the Tijuana–San Diego border, shifting the borderland again to make it more militarized and surveilled. In 1998, the Border Safety Initiative was established to collaborate with the Mexican government on preventing immigrants from crossing, as well as searching, rescuing, and identifying people along the US-Mexico border (US Customs and Border Protection, 2005).

The foundation stories of the border patrol are significant because they demonstrate how an organization develops from a questionable group of vigilantes into a highly funded organization across the United States, from Mexico to Canada. This new Migra's power now extends into our smartphones, in our database searches, through public institutes, and in other unexpected data gathering locales. La Migra has become a networked entity that is no longer just the physical border patrol in the Southwest. This new Migra is data-focused and data-driven, ubiquitous, and all-seeing. The new Migra is *networked* through Silicon Valley and *is* Silicon Valley.

It is challenging to bring to the forefront every player involved in the high value of immigrant data. So perhaps the best place to start is one company to rule them all: Palantir.

DEEP WEBS

Palantir

Palantir is an example of how this network expands into many marginalized communities: African Americans in New Orleans, the incarceration system, immigrant communities, and poor whites. Palantir's technology works across the nation, connecting law enforcement and the Department of Homeland Security to surveil communities of color. Palantir is *one* example of many systems working with ICE, border patrol, and the police on detentions and deportation. Other Silicon Valley partnerships built and strengthened off networked information abound: Clearview A.I., for example, established in 2017, already has more facial scans than the FBI (Haskins, 2019).

Palantir was established in 2004 by Facebook board member and PayPal cofounder Peter Thiel. After Thiel sold PayPal to eBay in 2002, he founded Clarium Capital, a global hedge fund, and then launched Palantir in 2004. In 2004 Thiel invested $500,000 for a 10.2 percent stake in Facebook; he sold his Facebook shares in 2012 for $1 billion. Thiel, who identifies as a staunch libertarian in what he considers liberal Silicon Valley, contributed $1.25 million to Donald Trump's presidential campaign (Waldman et al., 2018). Thiel was based out of San Francisco until 2018, when he moved his investment firm to Los Angeles (Waldman et al., 2018). Among Thiel's most famous conservative values is: "Most importantly, I no longer believe that freedom and democracy are compatible" (Naughton, 2016).

Thiel founded Palantir in the momentum of the September 11, 2001, terrorist attacks (9/11), choosing Alex Karp as his CEO. Thiel and Karp represent this new emerging Silicon Valley—Thiel being an alt-right Trump supporter and Karp representing the more socially liberal libertarian sympathies of the old Silicon Valley. Karp defended Palantir as the best surveillance tool for respecting citizens' civil liberties, compared to the "draconian policies" of the Patriot Act of post–September 11 America (Waldman et al., 2018). By 2013 Palantir was working with the CIA, the FBI, the National Security Agency, the Centers for Disease

Control and Prevention, the Marine Corps, the Air Force, Special Operations Command, West Point, and the IRS. Palantir built the Integrative Case Management (ICM) system for ICE, now used by law enforcement, and they built the Palantir Gotham software and support in 2013. However, until 2015 it was unknown to the public that Palantir was using predictive-policing data—analysis of data from the government, the finance sector, and legal research.

By 2015 Palantir had networked all their collaborators' databases for use by various law enforcement agencies. "Everyone from detectives to transit cops to homeland security officials uses Palantir at the LAPD" (Burns, 2015), which is now networked into ICE's databases, and there are many players involved in extracting, storing, and processing immigrant biodata. Palantir is an information management organization that catalogs networks of information and correlates that information into predictive policing.

The Department of Homeland Security aggregates data at their intelligence "fusion centers." Fusion centers collect data from many private and public entities without all users or targets knowing that their data is collected. That can include bank statements and credit card reports; daycare centers and preschools; colleges/universities and technical schools; health data (hospitals, mental health); veterinarians; jails, prisons, and court records; DMV records; court records such as property and mortgage deeds; and hospitality and lodging information from gaming, sports, amusement parks, cruise lines, hotels, motels, resorts, and conventions (Haskins, 2019).

One of those systems is Palantir's Gotham, a software program that holds a web of information profiles of communities of color. These databases of information work with ICE by networking information and profiling whom ICE should target in deportations next. A user manual from Gotham that *Vice* procured demonstrates how Gotham retrieves information: Police can run a name associated with a license plate and then bring up records that will give them a full report on where that driver has been at any given time. Police can also find personal information such as bank accounts, social media data, familial relationships, and that person's network (Haskins, 2019).

Government contracts developed the FALCON Search and Analysis (FALCON-SA), used by ICE as an analytical tool that visualizes data. A recent public document request by Motherboard (part of *Vice*) revealed Palantir's information collection targets, including banks, day care, jails, prisons, and even amusement parks, and demonstrates Palantir's visual training guide for how police use that software (Patel et al., 2019). ICM and FALCON-SA work as systems together to help ICE identify connections among so-called criminals and undocumented people through networked data. ICE has two major divisions, Homeland Security Investigations (HSI- ICM) and Enforcement and Removal Operations (ERO—stored in the Enforcement Integrated Database, EID), their case management systems. ERO uses Gotham-FALCON to decide who and where to raid based on algorithm predictions of networked connections between people.

They also use software to determine the future of those arrested; For example, the 'Risk Classification Assessment' software used by ICE determines whether an immigrant in custody should go to a detention center or can see a judge through algorithmic data. At the time this book was being written, Palantir was revising case management software, which allows immigrant agents to use regional, local, state, and federal databases to build profiles on immigrants, their friends, and family from private and public information (National Immigration Project et al., 2019, p. 3). Think of a Facebook built around anyone with a record and their extended networks.

The Cloud and Amazon

Data storage is a crucial part of the retrieval of and network creation with undocumented people's data. From 2008 to 2018, there was a significant shift in data storage from more spread-out data centers to the aggregated cloud. Before 2008, DHS had forty-three agency-specific data storage sites all over the United States. Data centers are undergoing migration to cloud storage under Amazon Web Services (AWS) to just two: Data Center 1, in Stennis, Mississippi, and Data Center 2, in Clarksville, Virginia. This project, called the Cloud First policy, was started in 2010 by the Obama administration's Federal Data Center Consolidation Services Initiative (National Immigration Project et al., 2019, p. 16).

In 2014, House Representatives introduced the Federal Information Technology Acquisition Reform Act (FITARA), which "expanded the powers of CIOs in federal agencies with regard to budgeting and IT acquisitions, and it called for data center consolidation and cost savings in federal IT operations, with quarterly reporting obligations" (National Immigration Project et al., 2019, p. 23). Simultaneously created with FITARA were the Cloud Computing Caucus, a congressional caucus, and the Cloud Computing Caucus Advisory Group, a lobbying group created by Amazon, Microsoft, and Dell (p. 23). Tech companies lobbied for

FITARA, and FITARA is *how* all of the data consolidations happened in cloud-first computing (National Immigration Project et al., 2019).

Along with Palantir, Amazon has a significant stake in immigrant data. AWS has become the leading provider of cloud storage, contracted under DHS. By 2019, AWS held the biometric data of 230 million unique identities, including 36.5 million face records, 2.8 million iris scans, and the rest fingerprint records (National Immigration Project et al., 2019, p. 5). Amazon acts as a federally authorized cloud service, holding 22 percent of federal authorization under the FedRAMP program and 62 percent of the highest-level authorizations to handle the data. Other cloud providers include (in order of FedRAMP Authorization): Microsoft, Akamai, Salesforce, Oracle, ServiceNow, Google, IBM, CGI, CSRA, General Dynamics, Granicus, SAP, MicroPact, Economic Systems, Zoom, and Adobe (p. 6).

Other tech companies that provide biotechnology to scan biometric information also house the data. For example, BI2, Biometric Intelligence, and Identification Technologies have worked with sheriffs along the border to provide iris-scanning tools like those they use in prisons on a free trial basis. Those technologies explicitly classify and search for the physical qualities of brown eyes and brown skin. Storage for BI2's data is in two law enforcement databases in Washington, DC, and Lafourche Parish, Louisiana, store BI2's information (George, 2017).

These surveillance tools are created to gather data, police communities of color, and predict *who* should be incarcerated or deported as an end goal. Since ICE has networked with some of the significant technology firms just discussed, it is vital to foreground how valuable the incarcerated are to prisons and jails.

END GAME: INCARCERATION AND DEPORTATION

The United States incarcerates people of color at an alarmingly high rate, with policing applied more heavily in their neighborhoods and communities and laws enforced with a swifter and heavier hand through the justice system. Institutional inequities and the school-to-prison pipeline also contribute to this uneven distribution of incarceration. While detention and incarceration is discursively discussed as a pipeline, it is also a life cycle, because recidivism rates in the United States are such that one person can cycle through these systems many times.

The United States has the largest and most profited prison and jail system globally, holding 20 percent of the world's imprisoned population (American Civil Liberties Union, 2022). Forty percent of incarcerated people are Black, while 13 percent of the total US population are Black. African Americans and Latinxs make up 32 percent of the population combined, but 56 percent of the incarcerated population as of 2015 (National Association for the Advancement of Colored People [NAACP], n.d.). Much of the research has demonstrated a disparity in how African Americans and Latinxs are policed in their neighborhood, charged disproportionately for nonviolent crimes such as drug possession, and sentenced to harsher sentences than whites. About 50 percent of federal drug cases are against Latinxs, yet they make up about 17 percent of the US population. Latinx men born in 2011 have a 1 in 6 chance of being incarcerated in their lifetime (Hernandez, 2019).

After incarceration, people with criminal records have their lives impacted in various ways: They often are not called back for jobs because of their criminal records, they cannot vote, cannot get housing, and have a high rate of infectious disease impacting their long-term health (NAACP, n.d.). Although incarcerated people have served their debt to society, in the traditional way that incarceration is presumed to lead to some form of justice, their record haunts them after release.

Incarceration has become a big business industry in and of itself. Between 1980 and 2015, the number of incarcerated people increased from 500,000 to 2.2 million (NAACP, n.d.); 2.7 percent of the adult population is under correctional supervision. Moreover, it is a business that targets young people of color early in their lives in the school-to-prison pipeline, where "policies that encourage police presence at schools, harsh tactics including physical restraint, and automatic punishments that result in suspensions and out-of-class time" (Elias, 2013). The school-to-prison pipeline results in African American K-12 students and children with disabilities overwhelmingly ending up suspended or expelled, heading to juvenile detention centers (Elias, 2013).

Immigrants in incarceration are not just in detention centers but often in jails and prisons. The process usually begins with local police and border patrol arresting undocumented people for different violations. More recently, they can be targeted through surveillance technologies and arrested based on being undocumented (Wise & Petras, 2018). ICE agents also arrest people at their homes, businesses, and, more recently, on their way to courthouses and at schools (Wise & Petras, 2018). There was a decline in immigrant arrests from 1.6 million in 2000 to 310,000 in 2017 (Wise & Petras, 2018). Some are deported with the expedited removal process—wherein they do not see a judge if they are apprehended within one hundred miles of the border and have been in the United States for two weeks or less.

ICE gives the Notice to Appear to people who have been arrested, and they are given ten days' notice before appearing in court. They are often sent to jails and prisons to wait out their court appearance and sentencing. After their sentencing and time spent in jails and prisons, they are sent to immigrant detention centers or contracted prisons. Some people can be released after arrest to pay bonds and if ICE agrees to those terms. A series of short hearings are held with federal immigration judges and an ICE attorney. During the master calendar hearing, an individual can file for staying in the United States based on three factors: asylum, cancellation of removal based on an individual's long-term residency, or marriage to a US citizen. During a merits hearing, the individual presents their arguments for staying in the country, and an ICE attorney is present. The immigration judge decides whether an individual will stay (Wise & Petras, 2018). During the merits hearing, the immigration judge may decide on an order of removal. Individuals can appeal or delay deportations to the Immigrations Appeals Board, of which there is only one in the country. When someone of Mexican nationality is deported, they are flown to the US border city and either walk or are bused across the border; people from Central America are flown directly by ICE air operations (Wise & Petras, 2018).

As in the early development of the border patrol, the Mexican government collaborates with USCIS and tech firms to collect and share data. For example, the US Customs and Border Protection's California Border Patrol signed the "Implementing Arrangement Regarding the Sharing of Border Crossing Data" with Mexico's National Migration Institute (INM) in 2017 to share and exchange biographic information of people crossing the border from the USA into Mexico (National Immigration Project et al., 2019). INM also shares border crossing data with CBP and biometric data with the US Department of State. Conversely, the DHS shares biometric data with the Mexican government through the Criminal History Information Sharing program (CHIS) through the EID case management system. CHIS also has data-sharing agreements with El Salvador, Guatemala, Honduras, the Dominican Republic, Jamaica, and the Bahamas (National Immigration Project et al., 2019). From 2017 to 2018, the Mexican government collected biometric data from over thirty thousand people in INM detention centers in Mexico City and along the southern Mexican border (p. 41).

Not only is incarceration and deportation an *end,* but it is also a cycle. Life after incarceration allows for only limited movement, housing, job opportunities, community ties, and life necessities, coupled with the already instated police state discussed earlier and the grim outlook that 76 percent of all inmates end up back in jail within five years (Hatcher, 2017). The United States has the highest recidivism rate in the world for prisons (Hatcher, 2017). That cycle includes being considered a criminal merely for existing in the country, in the case of undocumented immigrants. Deportation is now a part of their record, and they are entered into the cycle of incarceration. If they return to the United States, they return as a person with a record.

Incarcerated inmates are a valuable subject of the state and private industries. They are valued because it costs money to feed, house, and care for inmates but also because they are both tech customers and points of data. The cost of each incarcerated person varies from state to state. In Alabama, it costs $14,780 to care for inmates annually; but in New York, it is $69,355 per inmate annually (Mai & Subramanian, 2015). Within jails and prisons, incarcerated people are charged for using email, tablets, and other information technology (Austin, 2020). Jails and prisons control information and information technologies inside incarceration facilities, and they restrict incarcerated people's lines of communication with their friends, family, and correspondents. Information and communication can offer inmates a higher quality of life. A part of the systematic debilitating conditions of incarcerated people is to relocate them far from their homes geographically. Digital technologies become a necessary tool when prison and jail visits are no longer viable. Private, third-party vendors maintain technologies inside prisons and jails: they price tablets, telephone calls, and video calls at exorbitant fees for both the incarcerated and their loved ones (Austin, 2020). Biometric technologies are not excluded; for example, detention facilities use facial recognition, "developed by companies such as Face-First, which promises to geofence areas of the prison and send security alerts if people are to access unauthorized areas" (Austin, 2020, p. 3).

The new Migra network resituates undocumented people as subjects and identities built around data. If they do not have a name or social security number, they have a license plate number. If government-issued numbers cannot be associated with an individual, networked data can build up a data identity. These networks build and interconnect a silhouette of undocumented data bodies stripped of that individuals' qualitative identity. Through federally funded government contracts, this new Migra network is a big business industry that includes commercial and military industries and includes private and public everyday systems of information (such as the US Post Office's use of Clearview) that are assumed to be benign. One of the more baffling markers of this system (as discussed in the following chapters) is how enmeshed every resident's data is; *we* are inevitable members of this informant network. All data producers are welcomed and networked in the new Migra.

Such elaborate surveillance, control, and profit systems build prevailing and religious ideologies towards information technologies. The next chapter explores the ways in which all residents in the United States are connected to this network.

CHAPTER 4

The Good Citizen

Citizen Milieu

I didn't know I had any limits until I was in high school. Our parents don't talk about it because you're not supposed to talk about this with anyone—it's a secret that you're undocumented. My teachers wanted me to do dual enrollment with a college class, but the teacher told us we had to bring in our social security number the next day. I was afraid they would know because I wasn't coming to class. When you're sixteen everyone gets cars; at that time you needed a social to drive. It was one more thing I couldn't do. Or my friends would get jobs.

—Teresa

Whatever your citizenship status, it's *more likely than not* that you're networked into this data borders surveillance state. If you use the department of licensing, daycare facilities, shop at the grocery store, use academic databases for research, use the bank, or use social media, to name a few, you're probably networked into this Military-Industrial Startup Complex (National Immigration Project et al., 2019).

In this chapter I want to talk about the entrenched reliance of the data border on citizen milieu. *Citizen milieu* is the unknowing and often unwilling participation of US residents in this state of surveillance, deportation, and the centering of technological development around immigrants and the disembodied experience of invisible and ubiquitous data borderlands. It also involves the shifting use of information technologies to create the good citizen. By way of technology, digital media consumption, and sometimes simply existing in public spaces, citizens are always logged in as informants. The strength of the corroboration with the

Military-Industrial Startup Complex is the murky state of unknowing—we don't know when and how our data is being used to network into DHS and ICE. When the company that holds a government contract is private, they don't have to give us access to our own data (yet).[1] It's proprietary. But citizen milieu is not just the state of being unknowingly networked into surveillance, it is also a *part* of good citizenship: to be surveilled is marked as a part of one's civic duty. A hallmark of citizen milieu is the murky unknown of when and how our intimate data is being used for detention and deportation of others (Puar, 2007; Sweeney, 2020). Public service employees such as librarians and their patrons are uncertain whether they are participating in data gathering that leads to ICE surveillance and links into these major surveillance databases. That is the ubiquity of the data borderland and this specter of immigrant data that is always invisible and present.

So-called good citizens are those who are quantifiable, allowing data to drive their self-improvement. For citizenship, data moves in different directions; for good citizens, those who are documented or hailed as documented, this means being fully documentable as a consumer and citizen, lending one's data to correlating and building undocumented data bodies.

That goodness and belonging is now interlocked with data production, in opposition to undocumented subjects who lack data, whom Sun Ha-Hong has named the *good liberal subject:* "The good liberal subject is thus rearticulated as tech-savvy early adopters (who are willing to accept the relations of datafication before their validity is proved) and as rational, data-driven decision makers (who learn to privilege machinic sensibility above the human experience)" (2020, p. 7). Citizens are inscribed into the state and belonging by the production of data, but also by being intimately intertwined with the ICE network. If the villains of this data body milieu are those without data, then citizens are reinscribed as good guys through the production of data and the willingness to be correlated so as to act as a data point to build the immigrant's data profile.

Trying to understand and describe how citizens and immigrants are anchored into the ICE network is like feeling around in the dark for a

1. Many journalists, academics, and nonprofits have used the Freedom of Information Act (FOIA) to look at how this data is anchored into ICE. While FOIA can be used on government public agencies, it does not apply to private agencies, government contracts with private companies, or state governments.

needle in a haystack that may or may not exist, and this is part of the strength of the ICE–Silicon Valley network. We (civilians) don't know when our data is networked into surveillance and deportation processes. But we do know that we would have to boycott many systems to opt out of this network, such as social media platforms, government offices, and appearing in public. I use the first-person *I* and *we* often in this chapter as an aid to understanding. I, as a second-generation Mexican American, can be networked into this system at any given time, and so can you. This network is deeply personal, ubiquitous, and intentionally murky. I lead with the question: What does it mean when every data point is linked into the ICE network? What does it mean when most people are informers?

EVERYONE IS AN INFORMER: CITIZEN MILIEU

Who is a citizen?

There is a seeming dichotomy in this section between citizen and noncitizen that is in actuality not clear cut, and that should be acknowledged. But using the language of citizenship is valuable in naming the breakdowns in the citizen-immigrant binary. The lines between citizen and "undocumented person," or "alien" as the law labels them, are blurred when it comes to race, gender, sexuality, class, and nationality. A white, straight, cisgender undocumented person may be welcomed into the state's citizenship category with more ease and more rights than a Black or brown immigrant, an immigrant from the LGBTQ+ community, a lower-income immigrant, or an immigrant from a country that is not considered eligible for full citizenship in the United States.

Belonging to citizenship in the United States is an influx negotiation according to what benefits state formation. On belonging to citizenship, historian Mae Ngai says:

> Nevertheless, the line between alien and citizen is soft. . . Yet the promise of citizenship applies only to the *legal* alien, the lawfully present immigrant. The *illegal* immigrant has no right to be present let alone to embark on the path to citizenship. The *illegal* alien crosses a territorial boundary, but, once inside the nation, he or she stands at another juridical boundary. It is here, I suggest, that we might paradoxically locate the outermost point of exclusion from national membership. (2004, p. 6)

Illegal alienage is not a natural or fixed condition but one that is dependent on the historical moment and is fluid. Undocumented people have the possibility of becoming "legal," as permitted by that time and law, and legal immigrants can become illegal through various avenues as well. This fluidity and instability tell us more about the US national imagined identity. So there is legal and illegal belonging in our country according to the law, which is not fixed but changes depending on that point in US history.

Laws around citizenship can also target and produce those considered criminal and illegal. Lisa Cacho observes that poor people of color, undocumented immigrants, and gang members are termed "law breaking." "When law targets certain people for incarceration or deportation, it criminalizes those people of color who are always already most vulnerable and multiply marginalized" (Cacho, 2012, p. 4). Therefore, it is not only undocumented people without the correct citizenship papers who are cast as noncitizens but various populations deemed not to deserve the full protection of the law. Cacho explains that this population is ineligible for the legal recognition of full citizenship, full personhood by the law. Undocumented people are cast under what Cacho calls *social death*; they are dead to others in their eligibility for full rights and personhood. "As criminal by being, unlawful by presence, and illegal by status, *they do not have the option to be law abiding*" (Cacho, 2012, p. 8).

Citizenship is not always a spectrum of belonging, from undocumented to full citizen, but instead often an indication of who's in and who's out. At the time I'm writing this, elderly Asians are attacked on the street due to xenophobia (Song & Vázquez, 2021). This violence is long rooted in early anti-Asian laws such as the Chinese Exclusion Act and Japanese internment camps during World War II. After 9/11, anti-Muslim and anti-Arab sentiment led to a rise in hate crimes toward anyone with a veil, headscarf, or turban, including Sikhs. The citizenship status of the victims and targets of this xenophobia don't matter. They are targeted by white supremacy as outside of citizenship privilege based on long-rooted racism and orientalism. Law Professor Leti Volpp reminds us that, while on the surface President George W. Bush condemned the increasing hate crimes directed at Muslims, Arabs, and Middle Easterners, it was the not-so-subtle language around the "war on terror" and the actions of the US government that some US citizens took as their cues in hate violence. We see this same spectrum of attitudes toward Latinxs in public places speaking Spanish. Recently, video documentation has shown how Latinxs are attacked or assaulted while speaking Spanish in public (Gamboa, 2020). The citizenship status of those targeted is not accounted for in that moment of attack; it is an attitude that places that group on the outside of a roaming inclusivity of citizenship in the United States.

So what do I mean by citizen milieu? I mean that all US residents, at any given time, are networked into the ICE–Silicon Valley surveillance dragnet. However, that access to citizenship is often based on privilege, and it is a moving target depending on race, gender, sexuality, class, nationality, language, religion, and xenophobic contexts. The right type of citizen is welcomed through normativity and by being a willing participant of producing and correlating data.

How our data moves through networks like Palantir and tips off ICE for deportation is opaque. The back end of databases, algorithms, AI, and coding of those databases is proprietary to companies like RELX or Palantir. Rather, I would like to describe *how* it feels to try to touch the edges of it, while always uncertain as to *how* entrenched we really are.

THE STAKES IN CITIZENSHIP

Belonging to citizenship means having a data-minable information profile. Our data is the currency with which to access citizenship as identity, and it is socially stratified through those algorithms of oppression (Noble, 2018). Data and belonging in these networks, interdependent in incarceration and deportation, are also reliant on the constant *potential* for the exclusion and deportation of immigrants. Belonging and unbelonging of American citizenship kinship are done by way of racism and imagined kinships, but who is included in or excluded from citizenship is contextual. There is a rotating exclusion and inclusion to that kinship, depending on what benefits the state at different historical points of time. Inclusion of raced immigrants is just as telling about the structure of citizenship in the United States as exclusion.

Race Making

In the inclusion and exclusion of the social community in the United States, race as a criterion of inclusion and exclusion is in flux. As Ellen Wu (2013) says, race making is "the incessant work of creating racial categories, living with and within them, altering them, and even obliterating them when they no longer have social or political utility" (p. 7). Although race is *made,* it is naturalized, as Matthew Fry Jacobson (1998) observes, "in its ability to pass as a feature of the natural landscape" (p. 10). Jacobson writes on how European immigrants came to be identified with whiteness and white privilege, demonstrating that whiteness is built and maintained over time.

On race making in the United States, Sharmila Sen explains her personal experience of "getting race" when emigrating to America. Gaining or becoming raced is a prescribed identity due to racism, but it is also an ingrained ideology that is intrinsic to American cultural values. As Sen describes of her experience immigrating from India to the USA:

> Throughout this book I will use the odd formulation of getting race because I want to show you how I once perceived race as an alien object, a thing outside myself, a disease. I got race the way people get chicken pox. I also got race as one gets a pair of shoes or a cell phone. It was something new, something to be tried on for size, something to be used to communicate with others. In another register, I finally got race, in the idiomatic American sense of fully comprehending something. (Sen, 2018, p. xxvi)

While race in the United States is something that is made, it is also, as Isabel Wilkerson describes, a deeply ingrained caste system that is not easily movable but also central to US social infrastructure. While race is a social construct, it is the deepest level of structuring that passes itself as hypervisible and invisible in our daily lives. Wilkerson promotes the importance of stating that the United States has a caste system:

> A caste system is an artificial construction, a fixed and embedded ranking of human value that sets the presumed supremacy of one group against the presumed inferiority of other groups on the basis of ancestry and often immutable traits, traits that would be neutral in the abstract but are ascribed life-and-death meaning in a hierarchy favoring the dominant caste whose forebears designed it. A caste system uses rigid, often arbitrary boundaries to keep the ranked groupings apart, distinct from one another and in their assigned places. (Wilkerson, 2020, p. 17)

She notes that this caste system manipulates social systems as well as everyday behavior, finding that those in the upper-caste or more acceptable races exude "an inescapable certitude in bearing, demeanor, behavior, a visible expectation of centrality (p. 31).

Race and this imagined community do the work of excluding and including. As Wu demonstrates in relation to twentieth-century Japanese and Chinese Americans as the "model minority," race is a "historically contingent fabrication. Racial ideas do not appear out of nowhere and float around unmoored to social realities. They are consciously concocted and disseminated—if not always accepted without challenge—and are tied to structural developments" (2013, p. 8).

In post–World War II America, immigrants and nonwhite citizens were measured, tracked, and classified as belonging or excluded for the purposes of state formation. Wu demonstrates how Japanese Americans and Chinese Americans in the mid-twentieth century were nationally deemed the model minority through a series of intentional "consciously concocted and disseminated" manufactured decisions both outside of US authority and media, and inside. Those methods of assimilation never fully grant inclusion membership but are rather methods of racism that moved some Asian communities out of the American exclusion of material needs (such as housing or careers). For example, there was a push back against Japanese Americans participating in the Mexican American, Black, and Filipino trend of the zoot suit, a fashion statement that labeled young men as tough, and for Japanese American men and women to realign with American norms of gender and sexuality (Wu, 2013).

Methods of assimilation ultimately contribute to American exceptionalism in a time of decolonizing:

> In the end, the fashioning of Japanese and Chinese Americans first into assimilating Others and then definitively not-black model minorities did not only answer the question of Asian American social standing after Exclusion's end. It also worked to square the tension between the planetary spread of decolonization and the United States' designs to propagate its hegemony across islands and continents. Assimilating Others and model minorities performed an indispensable service for the imperative of narrating American exceptionalism to the nation and the world. (Wu, 2013, p. 9)

Asian Americans then, regarded as outsiders, were categorized as "accepted" foreigners through several choices in the postwar era; standardized into acceptable citizenship to an extent through intentional decisions and at the benefit of mythology around American liberal egalitarianism (Wu, 2013).

There is also a liminal state of citizenship, or as Volpp refers to it, a noncitizen state of exclusion that further closes the circle of inclusion. After 9/11, via the Patriot Act, Muslims, Arabs, and any person appearing Middle Eastern were often legally targeted for surveillance, loss of rights, and hate crimes. They were subjected to coercive interrogations by the Department of Justice, a more rigorous application of detention and deportation, and particularly targeted or removed without reason at airports and from airplanes (Volpp, 2002).

Post-9/11 surveillance culture and xenophobia have been top of mind for high-tech companies as they have developed their own surveillance tech. Just like the smart border wall, algorithmic surveillance is justified as less inhumane than the Patriot Act or a physical border wall. To vindicate algorithmic surveillance, Palantir's Thiel said that ". . . data mining is less repressive than the 'crazy abuses and draconian policies' proposed after September 11" (Waldman et al., 2018).

In reflection on the treatment of Muslims, Arabs, and Middle Eastern people in the United States and around the world after 9/11, the inclusion of citizenship was through the imagination of fellow members of that kinship: "citizenship as a process of interpellation starts from the perspective that power both subordinates and constitutes one as a subject" (Volpp, 2002, p. 1592). Citizenship is an identity. Beyond documents and legality, fellow citizens must be identified as such—they must be welcomed into the state by other residents. That process of recognition and welcoming, or interpellation, is the bridge to a full spectrum of citizenship benefits. It is dependent on the kinship of fellow citizens: "We function not just as agents of our own imaginings, but as the objects of others' exclusions" (p. 1585). Post 9/11, those "noncitizens"—those interviewed, deported, detained, and racially profiled into disappearance, are also the *disappeared,* along with those who were killed in the World Trade Center.[2] In the same way, we see the data consumer/citizen and the undocumented criminal/bad guy inextricably linked through the Military-Industrial Startup Complex's marketing on tech designed for ICE, border patrol, and police.

Post 9/11, the citizens' belonging to the state became dependent on the exclusion of Arabs, Middle Eastern people, and Muslims.[3] US kinship was built on their surveillance and profiling. In the current state of immigrant surveillance, detention, and deportation, kinship is increasingly built on networked data and the status of always being 'on'

2. Leti Volpp and Jasbir Puar have both noted that heterosexuality, heteronormativity, and gender roles were reinscribed and recirculated to double down on American imperialism in post-9/11 discourse around war and antiterrorist messaging. For example, the severe homophobic sexual abuse that occurred in various detention centers, most infamously at Abu Ghraib by the US Army, in parallel with expanding gay normative rights in the United States, is a way of re-engulfing American citizen exceptionalism. For more on this, see Puar's *Terrorist Assemblage: Homonationalism in Queer Times.*

3. Post 9/11, Mexicans and Central Americans, along with Latinxs perceived as a threat, were also included in the terror threat surveillance network, or "assemblage" as Jasbir Paur names it. US officials and residents imagined the border as a place where terrorists were crossing, so they restricted immigration and visas further. The Patriot Act only increased surveillance of Black and brown groups that were already not considered eligible for full citizenship benefits. See works by Puar and Lisa Cacho.

as a data informant. To be a citizen that is *identified* with such privileges means to be networked into this murky net of surveillance. We are never certain exactly where or when our data is moving in the ICE dragnet; it lives among the data of undocumented people in order to correlate their classification.

The Murky Intimacy of Information Bodies

For US residents, there are many data entries in the data border. For example, the Silicon Valley tech startup Clearview Facial Recognition, founded in 2017, has built the largest facial recognition system in the country. Over the years, Clearview AI has scaled back or increased their surveillance, depending on the status of human rights organization's lawsuits for data privacy violations (Ivanova, 2021). Despite being a young company, perhaps a defining feature of the Silicon Valley–based startup culture, Clearview Facial Recognition has a larger database of biodata than the FBI (Ivanova, 2021).

Citizen milieu lies in the murky unknowns of *who owns* the data and *where* our data *is* at any given moment. US citizens and undocumented people are now intimately connected through the uncertainty of where their data lives, as well as the value of data in and of itself. That is the ubiquity of this state. Data reaches far into the intimate corners of our lives. From the always listening smart TV to the uncertainty of if a hacker or NSA agent is watching through device cameras, our privacy is almost always accompanied by a possible other, unknown entity, AI or human, reaching out to us through data. The ubiquitous uncertainty of when and where this surveillance dragnet is happening *is* the state of data citizenship, an observation Puar made over a decade ago. "Control networks spiral through those who look, see, hear, gather, collect, analyze, target, scan, digitize, tell, e-mail, and tabulate, mixed in with those who are seen, heard, told, gathered, collected, targeted, and so on, the doers and those having something done to them concurrently or alternatively on no sides at all" (Puar, 2007, p. 154).

It is the perception of these all-encompassing surveillance networks as ubiquitous and infallible that circulates terror among the targets of those networks. This is the state of the data body, or information-creating bodies (Puar, 2007, p. 160). Citizens are networked into their bodies as information, and then correlated into data points. Our information/data bodies are out there, with billions of others, doing surveillance work when correlated together, and that data feeds back into spaces and access in citizens' and immigrants' physical bodies.

JPMorgan Chase & Co. learned this when, in 2009, they hired Palantir Technologies Inc. to monitor potential abuse of corporate assets. "Palantir's algorithm, for example, alerted the insider threat team when an employee started badging into work later than usual, a sign of potential disgruntlement" (Waldman et al., 2018). JPMorgan Chase's senior executives eventually learned that they too were being watched. Palantir used employees' financial documents, personal travel data, cellphone records, and social media postings as a data lab to correlate data and make inferences about those employees in their investigation on the matter. According to Waldman and colleagues (2018): "All human relations are a matter of record, ready to be revealed by a clever algorithm. Everyone is a spider gram now." Even the most elite of citizens with intersecting privileges can have their data working against or for them without their consent.

User Design: From the Private to the Public, from Home to Work

Citizens participate in these forms of surveillance in deeply personal and private ways, and they are designed with the user in mind, to play off consumers' levels of comfort as to what is familiar. The technologies are designed with the user in mind, to naturalize their integration into our private and public lives.

Let's take Amazon as an example. Amazon has become an intimate part of consumers' everyday lives through both their various technology devices and general consumer purchasing. The Amazon Echo and Alexa engage in consumers' data in everyday life. Critical information scholar Miriam Sweeney observes that Alexa, an anthropomorphized assistant, is designed as feminine to act as a buffer for acceptable surveillance in the home and in workplaces. An example is the use of virtual assistants in learning spaces such as libraries. In libraries, Amazon's Alexa and Echo are used for a variety of information services including event calendars, catalog searches, holds, and advocacy (Sweeney, 2020). Virtual assistants are designed to summon normativity around race, gender, sexuality, class, and citizenship, for the user to recognize as familiar and trustworthy.

Amazon is linked to ICE through their cloud-computing contract with DHS. While Palantir organizes the data through algorithmic inferences and interventions, Amazon Web Services hosts these massive databases, specifically a biometric database (Piven, 2019). Amazon's many digital services have become naturalized as a part of people's everyday lives. Amazon is in the home, on our bodies, and in public and private spaces through technology such as Prime, the Echo, Alexa, smartphones, and cloud services, to name a few. Perhaps one of the most prevalent ways that Amazon makes its way into users' private lives is through their virtual agent Alexa. The interface design is done with the user experience (UX) in mind. Alexa is an interface that takes on normative raced and gender qualities as a means of intimate surveillance. Sweeney (2020) argues that "digital assistant technologies mobilize beliefs about race, gender, and technology through interface design as a way to strategically cultivate UX, interpellate users as subjects, dismantle worker protections, and otherwise obscure (or 'smooth') vast intimate data capture projects. Tracing, and destabilizing, the role of anthropomorphic design in these systems is a necessary step for mapping the larger roles that digital assistants play in facilitating intimate data capture for the networked data environment" (p. 2).

Digital assistants are culturally coded as feminine through oral and textual speech patterns. They are also feminized through their domestic work as information service roles. These intelligent digital assistants are culturally coded to appear familiar *as* a white feminine voice such as Siri or Alexa (Sweeney, 2020). However, their normative interface design is the method by which they become easy to adapt to in the home and in workplaces: "anthropomorphism is but one strategy meant to cultivate user trust in the face of harmful data practices that rely on intimate surveillance" (p.7). Amazon is just one of many examples of how major information and technology products bump up against or overlap into government contracts with ICE through their web services. So the murky state here lies in not knowing if everyone's data resides together in Amazon's cloud services.

After digital technologies are naturalized into our everyday lives with user design in mind, we are by default opted in, or we opt in through cookies and add-ons, to this data mining dragnet.

RETROACTIVE DATA CONSCIOUSNESS

In 2011 I sat in Safiya Noble's office at the University of Illinois Urbana Champaign's African American Studies Department, and she told me some of the methods she was using to write what would become algorithms of oppression. She was looking at the ways in which the real estate app Zillow would label Black neighborhoods versus white neighborhoods, noting the racialized connotations Zillow applied if those neighborhoods were lower income, such as "low-rated schools." She breaks down the danger in the power of Zillow and data sets to name the value of different neighborhoods: "Never before has it been so easy to set a school rating in a digital real estate application such as Zillow.com to preclude the possibility of going to 'low rated' schools, using data that reflects the long history of separate but equal, underfunded schools in neighborhoods where African Americans and low-income people live" (Noble, 2018, p. 167). Fast-forward to November 2021, and Zillow has just announced that after allowing their AI to help people rampantly buy and sell massive amounts of housing throughout the COVID-19 pandemic, they will exit the housing purchasing market (Sherman, 2022).

Data surveillance privacy violations are experienced nonlinearly. For example, surveillance and policing might be happening for years before the public is aware of such violations. I call this phenomenon the *retroactive privacy violation consciousness*, in that consumers and surveillance targets may not know the extent to which information technologies were used until years after it has happened.

For example, in 2018, it was revealed that Palantir has been using predictive analytics, or "crime forecasting," in New Orleans since 2012. That system used networks between people, places, cars, weapons, addresses, social media posts, and other information to determine who to surveil through a similar system that mimicked stop-and-frisk policies. Predictive policing has also been used in communities such as Chicago, where police would use the same stop-and-frisk tactics following a list generated by an algorithm. This algorithm looks not only at those who have been arrested before but also at people within those people's social networks. The City of New Orleans–Palantir collaboration promised that this policing would result in social good—including engaging social services with the algorithmically identified target. However further investigation found that the program was not only ineffective but also served to racially profile networks of people (Patel et al., 2019).

Currently, retroactive legal action can take place long after a tech company, DHS, or ICE has violated data privacy. Learning that one's data is used for purposes of detention, deportation, and incarceration often happens as follows. A contract between ICE, DHS, border patrol, or the police will be given to a Silicon Valley company. Months or years later, a journalist, media outlet, scholar, or legal entity will use FOIA to receive information on the types of technology law enforcement personnel are using. That information goes public. Sometimes state authorities don't know the extent to which law enforcement uses tech and invades data privacy, and a lawsuit will be filed to stop that company from the data violation.

For example, Clearview AI has been found to collect biometric data in the form of facial recognition state by state. Once citizens are alerted to the use of this software and programs, they pursue legal action to get their data privacy back. These Silicon Valley military-industrial startups often operate outside of jurisdiction or without approval until they are caught. There is often little transparency or approval between law enforcement and the state (Congressional Research Service, 2020). Such was the case in Chicago, where Mayor Lori Lightfoot and the city didn't know that police were using Clearview AI software until that software contract was exposed by the *Chicago Sun-Times* (Schuba, 2020).

What stands out here is that *we* the public didn't know about this algorithmic policing and detention until 2018, though it had been happening since 2012. Thus, the *retroactive data consciousness* changes the ways in which consumers and surveillance targets experience time, as they are finding out about violations that occurred years previously. This phenomenon in user experience with IT warrants a need for more immediate awareness of potential privacy violations that we cannot foresee until they are egregiously out of control and eventually end up in congressional hearings.

There's something about the way algorithmic oppressions and power are experienced that is nonlinear, in that consumers and targets are not aware until years after that violation has occurred. Data sets are constructed and algorithmic decisions are made so quickly that we often come to socially understand the level of impact years after it's happened. But when looking at the bigger picture, we can see that critical information soothsayers were forewarning about these violations through their own research years before they happened.

In 2012, I sat at Dr. Miriam Sweeney's dissertation defense at the University of Illinois Urbana Champaign's iSchool and watched her discuss something called anthropomorphized virtual agents (AVAs). She analyzed the ways in which AVAs replicate human behavior in human-computer interaction and are also programmed by racialized and gendered tropes that underlie our larger social structures. This AVA in particular, Microsoft's Ms. Dewey, could be asked a question and respond with sexual innuendoes or racial stereotypes and tropes. Ms. Dewey was an early AVA that was discontinued quickly, but she was a precursor to the entrenched nature of AVAs to come. Fast-forward a few years later to find Siri and Alexa at every edge of consumers' private lives, the data privacy rights violations of those technologies always seemingly trailing their mass adoption. For example, in 2019 it was discovered that Amazon staff were indeed listening to the in-home audio recorded by the Amazon Echo (O'Flaherty, 2019). At the time of Sweeney's early work, I, like many technology consumers, thought that the AVA had problematic design functions, but those wouldn't be a part of my everyday life. What I could not see at the time was that Sweeney was forewarning about a technology whose problematic design features would be adopted into our homes and everyday lives. Sweeney was warning about the future, but consumers would retroactively know about these data violations years later.

Another mark of retroactive data consciousness is the way these companies and their public-facing data change. When I first started researching Anduril in 2019, their goals included predicting wildfires more quickly. Their website definitions of technology such as the Lattice operating system and their public-facing mission statements change over the years as the technological and surveillance landscape changes. The ground shifting out from under us is a hallmark of data borders: the way the public experiences borderland surveillance technologies' public-facing data does not have a stable core or linear timeline.

For example, in chapter 2 I discussed Anduril's mission as I had retrieved it in 2020:

> . . . consumer and commercial technology has outpaced the defense industry in both capability and cost. In addition, developments in AI, drones, and sensor fusion have thrived in the private sector resulting in the commoditization of the same advancements that will win and deter future conflicts.
>
> We are a team of experts from Oculus, Palantir, General Atomics, SpaceX, Tesla, and Google exploiting breakthroughs in consumer and commercial technology to evolve our defense capabilities radically. (Anduril, n.d.)

In 2022, that mission had changed to:

> The next generation of military technology will depend less on advances in shipbuilding and aircraft design than on advances in software engineering and computing. Unlike traditional defense contractors who focus primarily on hardware, Anduril's core system is Lattice OS, an autonomous sensemaking and command & control platform that serves as the core platform for our suite of capabilities. . . .

> Anduril is a defense products company. Unlike most defense companies, we don't wait for our customers to tell us what they need. We identify problems, privately fund our R&D and sell finished products off the shelf. Ideas are turned into deployed capabilities in months, not years. Saving the government and taxpayers money along the way. (Anduril, n.d.)

The ubiquitous and ephemeral cornerstone of Silicon Valley culture (one day a company may exist, and the next it may disappear, rebrand, or destabilize) shapes the public's experience of borderland technology branding, and comes in and out of focus in the public consciousness about their/our data rights.

CORRELATED COMMUNITY

Robert G. Reeve, a privacy tech worker, broke down *how* our data is shared and used on Twitter. Reeves demonstrates that our technology isn't necessarily listening to us but is correlating us as consumers. He discusses data correlation in a series of posts, worth the length to reproduce here:

> I'm back from a week at my mom's house and now I'm getting ads for her toothpaste brand, the brand I've been putting in my mouth for a week. We never talked about this brand or googled it or anything like that. As a privacy tech worker, let me explain why this is happening.
>
> First of all, your social media apps are not listening to you. This is a conspiracy theory. It's been debunked over and over again. But frankly they don't need to because everything else you give them unthinkingly is way cheaper and way more powerful.
>
> Your apps collect a ton of data from your phone. Your unique device ID. Your location. Your demographics. Weknowdis. Data aggregators pay to pull in data from EVERYWHERE. When I use my discount card at the grocery store? Every purchase? That's a dataset for sale.
>
> They can match my Harris Teeter purchases to my Twitter account because I gave both those companies my email address and phone number and I agreed to all that data-sharing when I accepted those terms of service and the privacy policy.
>
> Here's where it gets truly nuts, though. If my phone is regularly in the same GPS location as another phone, they take note of that. They start reconstructing the web of people I'm in regular contact with.
>
> The advertisers can cross-reference my interests and browsing history and purchase history to those around me. It starts showing ME different ads based on the people AROUND me. Family. Friends. Coworkers.
>
> It will serve me ads for things I DON'T WANT, but it knows someone I'm in regular contact with might want. To subliminally get me to start a conversation about, I don't know, fucking toothpaste. It never needed to listen to me for this. It's just comparing aggregated metadata. (Reeve, 2021)

Reeve demonstrates how powerful our data is in a community correlated to others. It is not our data *alone* that acts to target us as consumers, it is our data in relationship to others that makes consumers informers.

How, where, when is my data body?

Y(our) data as a product is worth more than market products such as shoes, clothes, or cars. In 2017, Wolfie Christl of Cracked Labs in Austria demonstrated *how* companies use personal data. Major conglomerates such as Google and Facebook "own" personal data, and other conglomerates *utilize* it for a price, to correlate that data for consumer advertising. Consumer data is gathered from such places as Facebook, Google, Apple, and credit reports.

Data from information such as social media profiles can be used to analyze and predict such nuances as personality traits, finances, insurance, and health (Christl, 2017). *Consumer data brokers* have had a long-standing collection of people's offline data and are collaborating and combining that with major online platforms including RELX (parent company of LexisNexis and Elsevier) to track and profile users. As he explains:

> [Data brokers] aggregate, combine, and trade massive amounts of information collected from diverse online and offline sources on entire populations. . . . The profiles that data brokers have on individuals include not only information about education, occupation, children, religion, ethnicity, political views, activities, interests and media usage, but also about someone's online behaviors such as web searches. Additionally, they collect data about purchases, credit card usage, income and loans, banking and insurance policies, property and vehicle ownership, and a variety of other data types (Christl, 2017).

Data brokers such as Acxiom and Oracle collect individual's data on- and offline without their knowledge or permission and build data profiles, with scores, that can predict an individual's behavior. At the time of Cracked Lab's 2017 report on the topic, Acxiom had data profiles on 700 million people, and Oracle had data profiles on 1 billion mobile users and 1.9 billion website visitors, with a total of 5 billion unique consumer IDs (Christl, 2017). Selling user data requires data management platforms, businesses that combine and link their own data on consumers with digital profiles, using correlation to combine and link data for digital tracking (Christl, 2017). These data profiles lead to tangible results in people's everyday lives: "Every click, swipe, like, share, or purchase might automatically influence how someone is treated as a customer, how long someone has to wait when calling a phone hotline, or whether someone is excluded from marketing efforts or services at all" (Christl, 2017).

The stakes in this data dragnet are high, and they open and close invisible doors. Axciom's annual revenue as of the 2018 fiscal year was $917 million (Acxiom, 2018). Citizens and immigrants are a highly valued data brokerage product in the United States. It's difficult to put a price on each individual as a data point. In their FAQ, Acxiom's website goes with the gaslighting approach: How much is your data worth? Not much, they claim (Donovan, 2015). But data brokering is a $200 billion industry that puts a price on data lists to sell to companies, with $89 the average value of one email address over time, and companies do not let consumers fully opt out of their data being brokered (Jenkings, 2012). In one instance, a company sold lists of one thousand people each with various medical conditions and experiences including rape, alcoholism, and erectile dysfunction, for $79 (Hill, 2013). Citizens are consumable data points, and those data points can be networked into ICE surveillance as well.

Let me present myself as an example of how I am networked into the ICE surveillance dragnet daily:

I wake up and drive my daughter to *day care* with my car, which is registered with the *state government.* I am a professor and I teach information science, so I teach my students *databases* housed under RELX and Thomson Reuters, to prepare them for their careers as librarians, where they will go on to use and teach these databases to their patrons. I live in Rhode Island, which is within one hundred miles from the ocean (a *border*), so US Customs and Border Patrol, under the hundred-mile border zone, can claim power of warrantless searching including electronic devices at the "border," which incidentally covers two-thirds of the US population (American Civiil Liberties Union, n.d.). I'm using *Microsoft Office* to write this book and conduct most of my written work, a company that provides the Office 365 Multi-Tenant & Supporting Services to the DHS's FedRAMP authorization (National Immigration Project et al., 2019). My workplace uses *PlumX,* owned by *Elsevier*, a data visualizing system to organize our open-source institutional repository. When I requested my data profile from LexisNexis Risk Solutions, pursuant to Section 609 of the Fair Credit Reporting Act (15 U.S.C 1681g), I received sixty-two pages of my own data that

LexisNexis has compiled, including past employment, past housing, past inquiries into my credit from banks or creditors, my driving record including a 2018 parking violation, previously attended schools, my parents and former and present partners' data (LexisNexis, personal communication, July 7, 2022).[4]

Beyond those technologies that we use in our everyday lives, there are often surprising entrances into ICE data surveillance. The library is one of those workplaces and public spaces where this surveillance technology is adopted and where the edges of these data borders extend. The library has become an unlikely space where unknowing participants touch the edges of this policed data borderland.

THE LIBRARY: A MICROCOSM OF EVERYTHING

For decades, IT and data privacy activists warned us about the dangerous of a few large companies holding ownership over access to scholarly materials. In the early days of public internet access, information scholars and activists expressed concerns over what James Boyle (2003) named the "second enclosure movement," which refers to the proliferation of intellectual property. Boyle argued that copyrights over ideas can backfire and prevent further innovation, interfering with the prosperity of the public. Boyle's warning, written two decades ago, can now be seen as a foreshadowing of data privacy to come. Through the 1990s and following decade, information and technology activists, at times called "hacktivists," warned about the move to privatize scholarly thought. From technologists to librarians, the concern was often voiced that as scholarship was digitized, it would become more closed to access by the public.

Technology inventor, hacker, and activist Aaron Swartz was one of these voices. Swartz helped develop such innovations as the RSS feed and the Markdown publishing format, and he helped to organize Creative Commons. In 2010, Swartz connected to the MIT network and used JSTOR to download a large number of academic journal articles. Swartz wrote in the *Guerilla Open Access Manifesto:* "The world's

4. To reach out and touch this data border, as an example of US residents' anchoring into systems that collaborate with ICE, people with social security numbers and/or driver's license numbers who feel comfortable doing so can request their data profiles with LexisNexis Risk Solutions online request form at https://consumer.risk.lexisnexis.com/request.

entire scientific . . . heritage . . . is increasingly being digitized and locked up by a handful of private corporations. . . . The Open Access Movement has fought valiantly to ensure that scientists do not sign their copyrights away but instead ensure their work is published on the Internet, under terms that allow anyone to access it" (Swartz, 2008).

Whereas state prosecutors dropped the charges against Swartz of breaking and entering, grand larceny, and unauthorized access to a computer network, federal prosecutors filed nine felony counts, which could have led to fifty years of imprisonment and $1 million in fines. Prosecutors negotiated with his attorney for him to plead guilty to a thirteen-count indictment. On January 11, 2013, Aaron Swartz was found dead by suicide. The federal prosecutors were chided by the public for such extreme disciplinary actions, but the incident would be a forewarning of the high value of intellectual property.

Coinciding with Swartz's and other hacktivists' concerns, librarians have long voiced and protested around their concern for an individual's right to data privacy *and* parallel to that, the right of the public to have access to scholarly materials. As information has gone digital, librarians have experienced vendor prices skyrocketing and have struggled to secure access to journals and databases for their institutions (Lamdan, 2019).

Library vendors are companies that own most databases that libraries and their patrons access. For example, if a college student or K-12 student must write a report, they may go to the library and ask for help conducting research. Those librarians may advise them to search keywords to narrow in on scholarly research on the topic. To have these information resources, librarians purchase access to large databases, which house smaller databases and access to scholarly journals. Over the decades, data service corporations have become some of the most profitable tech companies in the industry, without the public scrutiny of Silicon Valley infamy (Lamdan, 2019). This information monopoly stronghold has led to a few publishers owning most of the access to scholarly materials, news sources, and more in library collections.

Recently, what seems both impossible and inevitable happened. Thomas Reuters and RELX Group (Elsevier, Westlaw, and Lexis Nexis) gave access to all their patrons' data through government contracts. Since 2013, ICE's Fugitive Operations Support Center has contracted with LexisNexis for access to their databases (Lamdan, 2019). The way it works is this:

> ICE pays RELX Groups and Thomson Reuters millions of dollars for the personal data it needs to fuel its big data policing program. Thomson Reuters supplies the data used in Palantir's controversial FALCON program, which fuses together a multitude of databases full of personal data to help ICE officers track immigrant targets and locate them during raids. LexisNexis provides ICE data that is "mission critical" to the agency's program tracking immigrants and conducting raids at peoples' homes and workplaces. (Lamdan, 2019)

Just as Amazon was once just a bookseller and is now the largest cloud service host for ICE data, these companies, once publishers, are now data brokers.

In 2019, lawyers, students, scholars, Mijente, and the Immigrant Defense Project wrote letters and petitions calling for RELX and Thomson Reuters (owner of the Westlaw legal search tool) to end their government contracts with ICE. Those activists pointed to the ethical bind of immigrant rights lawyers relying on those legal databases for their clients (Currier, 2019). Every law firm relies on Thomson Reuters and RELX. LexisNexis maintains that they do not use information from legal research tools, asserting that customer data and quieries were never sold into any product databases (Currier, 2019). They insist that the same digital products that consumers and customers use are networked into ICE, but that there is no overlap from customer data to ICE targets: "Spokespeople from Lexis Nexis and Thomson Reuters, asked about their work with ICE, both emphasized helping law enforcement with 'public safety.' Services provided to ICE 'by Thomson Reuters are explicitly in support of its work on active criminal investigations and priority in cases involving threats to national security or public safety'" (Currier, 2019). As in many instances of tech development for borders, detention, and deportation, the specter of the undocumented criminal overrules data privacy rights concerns. As with Amazon, the attitude is that we'll just have to trust those companies that the so-called criminal data and the so-called consumer data are separate in surveillance.

Stuck in the Citizen Milieu

When City University of New York Professor of Law and former librarian Sarah Lamdan discovered that ICE was already in the library stacks, she began to push back. She realized that the very databases that she taught and promoted to students, scholars, and lawyers were those that held government contracts with ICE. She asked herself about the ethical ramifications of this situation. What's more, she found that RELX and Thomson Reuters were using data as data brokers—a fact that many patrons and librarians were unaware of. She wrote:

> These companies seem to be warehousing and selling all of the datafied information they can grab. Even though we still treat each data market separately (for instance, academics criticizing journal costs don't see the similar problems in legal information services) they have become overlapping circles in the same Venn Diagram, intervening in several corporate data powerhouses at the Diagram's center. Personal data brokering, legal information, pay walling, and high academic research costs are all disparate data problems caused by the same few businesses. (Lamdan, 2022, p. 2)

RELX and Thomson Reuters are data monopolies, but not held to the same standards of accountability as those much more public-facing companies like Google, Facebook, or Apple. Lamdan began to educate other librarians about the deeper ties to these vendors. Lamdan and a fellow librarian posted a blog post on this very topic to her professional library association, the American Association of Law Libraries. That post was very quickly removed.

What Lamdan learned over the course of years was that librarians, libraries, library associations, and their respective institutes were already so intertwined with these vendor relationships (monetarily and in kinship) that to openly challenge them would lead to quiet censorship. Those professional relationships, like human-computer interfaces such as Alexa, are intentionally built around intimacy. Lamdan saw the buy-in between librarians and data companies over the years at library conferences, but it was a fraught, one-sided relationship: "Among librarians and scholars, RELX and Thomson Reuters aren't known as benevolent—they're known as bullies. The companies exploit scholars' free labor and then steal the fruits of their research, paywalling their work so that only subscribers can see it" (Lamdan, 2022, p. 5). Because the thing is, over the years the *exact* situations that hacktivists such as Aaron Swartz had voiced concern over were coming about. LexisNexis and Elsevier were constantly raising their prices for access to scholarly materials, law records, archives, and more, and librarians were often at their mercy to either pay those high prices or walk away and lose their patrons' access to irreplicable materials.

As Lamdan spoke out against government contracts between RELX, Thomson Reuters, and ICE, LexisNexis representatives took measures to push back. She received pushback from LexisNexis when she openly spoke about ICE entanglements, in the form of phone calls from Lexis's regional product managers and Lexis reaching out through her students (S. Lamdan and M. Aizeki, personal communication, 2021). Lamdan found that a perfect storm of information inequity has come together over the years of hoarding data and then giving that data to ICE, and she calls those major companies *data cartels:* "The data analytics companies are acting like cartels, taking over markets for different categories of information and exploiting their stronghold on multiple information sectors to amass more power and profit" (Lamdan, 2022, p. 2).

When Lamdan saw threads between her data, her work, and ICE detention, she pulled on them harder. What she found is that our data is so entangled and held hostage by these tech companies that any challenge was pursued with hostility. These companies are monopolies and monoliths. But the veneer of their relationship to consumers is that of friendly data companies with a concern for user design. As Lamdan continued to challenge those data cartels, she was quietly moved away from relationships with vendors. As she explained in a webinar on ICE surveillance through digital library tools, she was also told by those companies that she could not teach her students about the government contractor's relationships to search engines (S. Lamdan and M. Aizeki, personal communication, April 8, 2021).

Our data bodies live together, out there, as valuable products to data brokers, tech companies, and ICE. These companies insist that their *customer* data is not mined for ICE privileges, cutting a clear dichotomy between scholarly use of database searching and undocumented peoples' data. This is not only ethically impossible for lawyers, law students, librarians, and law practices to navigate, but it positions undocumented people as not possibly belonging to the scholarly audience that uses those databases. They are placed on the surveillance side of RELX and Thomson Reuters.

For customers regarded as residing on the "good" side of citizenship, those whom RELX insists are not surveilled, the friendly relationship between consumer and company is thin, as Lamdan's experience demonstrates. For those with privilege, that data may be excised to approve a loan, approve treatment for medical care, or determine if someone can keep their job (Eubanks, 2017; O'Neil, 2016). Other librarians have also reached out to these data brokers to get a look into the backend of where our data lives. When library and information studies scholar Shea Swauger requested his own data, Thomson Reuters sent him forty-one pages of personal information: "Thomson Reuters attached a waiver to the data absolving the company of responsibility for the errors (2019). Neither RELX Group or Thomson Reuters ensures that they're providing comprehensive, complete, or accurate data. Thomson Reuters admits that your data could be mixed up with other peoples' data. The companies don't recommend depending on their data for any purpose" (Lamdan, 2022). What stands out here is that Thomson Reuters cannot guarantee that each person's 'data body' is separated from others. Our data bodies are interrelated and dependent on each other to correlate and inform.

CONCLUSION

Here is the bottom line for citizen milieu: our data bodies live, out there, with others and without context. By way of commercial data brokering, our data bodies engage in the ICE–Silicon Valley network of detention, incarceration, and deportation. The cloaked nature of where our data lives and what it engages with *is* the state of power held by data brokers.

Lamdan's experience with data cartels is unique and alarming. As data consumers we rarely feel empowered to push back on having ownership over our own data. It's easy to feel lost in the third-party add-ons, accepting cookies and extensions, losing track of the many companies that hold government contracts with ICE. But what it demonstrates is just *how* engrained we are in this state of interconnection with ICE, detention, and deportation, and the high value of our data as a living organism among data brokers. What it signals is that while there is often a user-friendly veneer on the public face of data brokering companies, when a consumer pushes on that network, it bites back.

I began this book with my own descriptions of the borderlands in Southern California. The sights, smells, sounds. The *feeling*. But as we continue through the journey of this book, that anchor loosens its hold. Infrastructure and networks spread out throughout the United States. Borders lose context. People lose context.

When I first read Sarah Lamdan's work, I had a visceral response to realizing my everyday work was anchored into the ICE network. I remembered the time my mom was questioned by border patrol and had to repeat "US citizen" to all their questions. I thought about when two police officers were harassing migrants looking in our neighborhood trash for cans to recycle. When my mom ran up to the patrol officers and told them to leave them alone, one officer laughed in her face.

In this state of citizen milieu, the surveillance network, those detention centers, the ICE raids, the borderland surveillance are always already at our workplaces, in the home, and in public spaces. Critical information theorist and librarian Jeanie Austin asked in a presentation called "Identifying 'Trap Doors'" at the 4S annual meeting in New Orleans in 2019: "When do you lose control of yourself as a point of information?" If you're in the United States, it's more likely than not that you are touching the edges of those data borders, if just as a correlation point.

In this chapter, I described the state of being networked into this ICE–Silicon Valley data dragnet. In the book's conclusion, I will speak about methods of agency against this network. I continue to circle around the *how-did-we-get-here* question in chapter 5 by looking at the larger values held in our society around technology, and the rifts among values within Silicon Valley. Data borders are built and sold by way of story structures that engage the surveilled, consumers, employees, and CEOs of these Silicon Valley companies. Next I look at how storytelling and the narrative of the hero's story arc frames many of the tech companies that collaborate with ICE, engaging their own workers as players on the border.

CHAPTER 5

The Stories We Tell

Storytelling for Data Borders

Back then immigration was on their *motos*, in the helicopter, and on the horses. But now they are watching everything with technology.
—Rosa

WIRED FOR STORY

Beyond the infrastructure of immigrant data surveillance and belonging as citizens, we find ourselves entrenched in the *story* of citizenship, immigration, and (un)belonging; storytelling is the skeleton onto which borderland information technologies are developed. "We are wired for story," Lisa Cron writes about this human condition (Cron, 2012). The biological function of story has become a focal point in self-development culture and a motivator for corporate interests (Brown, 2015). While the human instinct for shaping story is to understand personal consciousness and how we relate to each other in community, storytelling is also a function for capital gain: "The idea of storytelling has become ubiquitous. It's a platform for everything from creative movements to marketing strategies" (Brown, 2015). When selling a product, marketing departments and entire corporations recognize the power of story in our biological and social structures.

Stories and storytelling hold power. They resonate with an evolutionary function within our brains and body that reaches back to survival basics (Cron, 2012; Brown, 2015). Anduril, Palantir, and other tech companies have built immigrant data surveillance technologies around storytelling because drawing on a story is an innate and deep-rooted instinct within the social fabric that builds communities. Storytelling, myths, and legends are a tool for Silicon Valley companies to tap into a a primordial biological function that positions the players on this stage—citizens, immigrants, the government, and tech companies—in a clean good-versus-evil binary.

This chapter focuses on how story propagates technologies that surveil and are designed around immigrants and immigration, and how storytelling and stories reinforce the roles of immigrants and citizens. However, a more significant concern here is not just that humans are wired for story but also that we can be shaped by story lines running through this emerging data borders state (Cron, 2012; Gottschall, 2013).

We need story to make sense of and know how to react to situations. Story also matters on a societal level; our culture uses stories to make sense of norms and values. Borderland tech companies use the power of story and myth to realign citizens and border tech players as the heroes or good guys of surveillance technologies, and undocumented immigrants as the villains. In this chapter, I want to explicate the stories and myths constructed by Silicon Valley companies. There are multiple ways in which story mobilizes the development of these technologies. The overarching story is one in which the heroes are called to action with hybrid military-commercial IT to defeat the bad guys. The borderland and Latinx immigrants are the threat used to promote this new industry. As those heroes encounter resistance in the form of protests and criticism, they find themselves changed in building military-commercial products that serve ICE. They must leave what they previously knew to build something new.

I first look at the power of stories in individuals and our social structures. There are chemical reactions that respond when we engage in story, and there is a significant buy-in to being part of a story's plot. Then I want to look at the story line occurring for the larger immigrant surveillance companies related to the heroes' journey: the world's rules are laid out, the hero ventures out into new territory, and the heroes come back changed.

On the evolution of storytelling media through technology, American literary scholar Jonathan Gottschall notes that "the technology of storytelling has evolved from oral tales to clay tablets, to hand-lettered manuscripts, to printed books, to movies, televisions, Kindles, and iPhones" (2013, p. 186). His conclusions about where storytelling was going are clear: to the virtual, digital world. Humans have evolved to use, need, and at times misuse stories. Digital technologies are designed around such a biological instinct: ". . . as digital technology evolved, our stories—ubiquitous, immersive, interactive—may become dangerously attractive. The real threat isn't that the story will fade out of human life in the future; it's that story will take it over completely" (Gottschall, 2013, p. 198).

This chapter illuminates how story is engaged and engages the user in this state of data body milieu. However, first I want to establish the importance of story, individually and socially.

The Power of Story

Feelings, and not rational thinking, are running the show in human decision making and subsequent actions, and it is *story* that invokes those feelings: "we think in story, because story provides a context for the facts, so we can make sense of them. It's this very subjective process that gives the meaning, triggering the motion, the feeling, then silently guides our every action. . . . [S]tory rules your life" (Cron, 2014).

Story has a biological function: our brains use stories to tell us how to survive. For example, if I see a bear, the story in my brain tells me, "that bear is dangerous, react." Moreover, when we listen to a story, we receive a chemical reward in our brain for successfully protecting ourselves: "It is the surge of the neurotransmitter dopamine that's triggered by the intense curiosity that an effective story always instantly engenders. It's your brain's way of rewarding you to follow the story to find out how the story ends. . . . We don't turn to story to escape reality, we turn to story to navigate reality" (Cron, 2014). The brain learns from feelings subjectively. So when I survived the encounter with the bear (in that story), my brain gave me a chemical reward, and I *learned* that *that* story was correct. The bear was dangerous; I responded by protecting myself (fought, fled, or froze), and I learned from that experience/narrative/story. Stories in our brain are running the show, not facts and objectivity. We change our behavior based on storytelling and stories, not logic or reasoning.

Story dictates our responses in our brain by drawing on the feelings invoked. Gottschall investigates what he calls *the storytelling animal,* the power of stories in humans. When humans read, view, or listen to a story, we are experiencing the thing in our brain as we watch it. For example, if we are watching a scary movie and the monster is chasing the protagonist, the part of our brain is lit up as if we, the viewer, are being chased by a monster: "Story is so powerful for us . . . because on the neurological level, whatever is happening, on the page or on the stage, isn't just happening to them, it's happening to us as well. We know it's fake . . . but that doesn't stop unconscious parts of the brain from knowing it as well" (Gottschall, 2014). When we watch movies and shows of characters, we move in empathy along with them, and it can move society toward change or align a society with the *moral* value of that story (Gottschall, 2014).

The term *story* refers to books, television shows, movies, and other media, as well as the stories we use in our minds in everyday engagements, like the "see bear, react, survive" story. Brain researcher Paul Zak has found that hearing/viewing/reading/believing a story releases cortisol and oxytocin in the brain, chemicals that help humans connect, empathize, and make meaning (Brown, 2015). So we humans must make meaning out of our experiences, and we *have to* make up a story.

We make decisions based on previous stories that helped us in the past and help us navigate reality in the present and future. But a caution: that chemical reward, dopamine, will be triggered whether we were wrong or right about a story. It is the unknown, the unexpected, that story helps us anticipate. Stories feel good for the same reason that food tastes good because without it, we couldn't survive . . ." (Cron, 2014). According to neurologist Robert Burton, a *pattern-recognition reward* is when humans make assumptions to fill in parts of the story that are unknown (quoted in Campbell, 2017). As social scientist Brené Brown explains Burton's research, a dopamine reward is earned in completing a story or pattern: ". . . Stories are patterns. The brain recognizes the familiar beginning-middle-end structure of a story and rewards us for clearing up the ambiguity" (Brown, 2015).

Our brains have multiple chemical rewards released when we have completed a story in our minds, whether that story is correct or not. Brown (2015) notes that the danger that happens in pattern-recognition reward is that the brain begins to make up stories to make sense of the world. The brain fabricates stories to complete the loop for meaning making: "The storytelling mind is allergic to uncertainty, randomness, and coincidence. It is addicted to meaning. . . . In short, the storytelling mind is a factory that churns out true stories when it can but will manufacture lies when it can't" (Gottschall, 2013, p. 103). Conspiracies are the mind's way of making meaning out of ambiguity.

We can see this in stories around immigration, immigrants, and the border, mainly those involving political rhetoric. What happens when the jobless rates go up and more people find themselves struggling financially? They draw on a long-established story that immigrants are taking their jobs. Coinciding with that story are many other stories around different immigrants: They're lazy, they're hyper-reproductive, they're criminals, they're taking our jobs, they're smuggling in drugs. Various responses to these powerful stories come from anti-immigrant policy and surveillance technologies that promise to prevent further immigration or lead to deportation. In other words, they promise a *future* or even *futurity* without certain immigrants. Borderland tech companies fill in the story arc and tap into that evolutionary instinct that motivates us to protect ourselves: immigrants are in this country to [fill in the blank], and tech companies are here to save us by promising a future without them.

Story also serves a social function that is intertwined with survival.

The Social Body, Imagined

Individuals engage in stories daily. What is the role of story and myth in our society? Humans are innately social creatures, and building community also creates a physiological reaction. Oxytocin is released when prosocial behavior happens, through such engagements as sex, taking care of one's offspring, or socializing with friends and family. "In humans the brain regions associated with emotions and social behaviors—namely, the amygdala, hypothalamus, subgenual cortex, and olfactory bulb—are densely lined with oxytocin receptors" (Zak, 2012, p. 38). Social engagements that generate trust engage what Paul Zak calls the moral molecule. Positive social stimuli release oxytocin, and dopamine and serotonin are released to reduce anxiety (Zak, 2012, p. 38). Socially, humans engage with stories in all parts of our lives: "Story—sacred and profane—is perhaps *the* main cohering force in human life. A society is composed of factious people with different personalities, goals, and agendas. What connects us beyond our kinship ties? Story. . . . Story is the counterforce to social disorder, the tendency of things to fall apart. Story is the center without which the rest cannot hold" (Gottschall, 2013, p. 138).

Foraging societies evolved toward groups rather than a sense of the self. Our contemporary moral selves are evolved out of that early sense of the group:

> When a person does something for another person—a pro-social act, as it's called—they are rewarded not only by a group approval but also by an increase of dopamine and other pleasurable hormones in their blood. In

> addition, group cooperation triggers higher levels of oxytocin. . . . Oxytocin creates a feedback loop of good-feeling and group loyalty that ultimately leads members to 'self-sacrifice to promote group welfare,' in the words of one study. (Junger, 2016, p. 29)

Story works on the communal level as a social glue. However, social glue also relies on stories around outsiders that make insiders possible. Immigrants and immigration have a long-established "story" in the early formations of the United States; they are part of an often-false narrative that has built American into its sense of self. We often hear that "we are a country of immigrants," a sentiment that disregards the colonial and slavery past but works as a *story* that lends itself to the pull-ourselves-up-by-our-bootstraps national imagination. Nevertheless, inclusion into state and national identity is dependent on *which* undocumented immigrant, at *what* point in history, and from *which* locale, ethnicity, and race.

This is the "imagined community" named by Benedict Anderson. A nation that uses story, or imagination, to develop its sense of self, belonging, inclusion, and exclusion: "It is *imagined* because the members of even the smallest nation will never know most of their fellow-members, meet them, or even hear of them, yet in the minds of each lives the image of their communion" (Anderson, 1991, p. 6). "Imagine how citizens imagine their comradery under nationalism and are willing to sacrifice themselves for a larger purpose. The story, the imagined part, is the comradery among citizens" (Anderson, 1991).

Undocumented immigrants, which we can also think of with Mae Ngai's term *impossible subjects*, "are at once welcome and unwelcome: they are woven into the economic fabric of the nation, but as labor that is cheap and disposable" (Ngai, 2004, p. 2). Undocumented immigrants may be considered a *caste*. They inhabit an informal status in the United States that cannot be denied but is not legitimated, and a caste often belongs to an ethnoracial minority group (Ngai, 2004). The restrictions and exclusion of *some* immigrants is also a part of this American story of belonging and state building via immigration policy: "immigration policy is constitutive of Americans' understanding of national membership and citizenship, drawing lines of inclusion and exclusion that articulate a desired composition—imagined if not necessarily realized—of the nation" (Ngai, 2004, p. 5). Undocumented immigrants are presented as a threat to citizenship and tighten the story of the citizen's sense of belonging to their imagined community, the nation.

Part of building this imagined community is the making of race. Story structure around data borders and data body milieu also do this work of race making through intensified storytelling.

STORY STRUCTURE AND THE HERO'S JOURNEY

Story structure happens in multiple ways: as a chemical reaction in our brains that are wired for story, and also on a social and national level that creates an imagined sense of belonging on a group scale, not without social hierarchies of race, gender, class, and sexuality. As Wilkerson notes, as a human division, race/caste "surpasses all others—even gender—in intensity and subordination (2020, p. 20). Story structure for borderland technology works for citizen inclusion and exclusion, building state formation (Chaar López, 2021) and determining desired and undesirable types of Latinx immigrants. One question to keep in mind is, how do these story structures continue or shift the fabrications of race, belonging, exclusion, and Latinidad in the United States? In this chapter I'm concerned with the many groups involved in this storytelling of data borders. From CEOs to tech workers in the lowest ranks of these companies, many are called upon to cosplay the hero in these data border stories.

In his seminal book *Hero with a Thousand Faces,* Joseph Campbell (1949) outlines how the hero's journey plays out in three parts. In Act I, the world's rules are established, and the hero is called to an adventure: we come to understand how that world works, even if it is a fantasy. At the end of Act I, the hero must respond to the call to action (Reagon & Brown, 2020). In Act II of storytelling and the heroes' journey, the hero attempts to find a comfortable way to solve the problem and realizes that they cannot solve that problem without leaving their own world where they are comfortable, thus forced to venture outside of their comfort zone/world. Finally, in Act III, the protagonist learns a lesson, proves it has been learned, and returns to their world, changed (Reagon & Brown, 2020). Because these military-commercial hybrid Silicon Valley companies draw on classic hero-villain binaries and hero journey story structures, I will also use that structure in this chapter.

In the following sections, I look at the story line playing out around Silicon Valley's technology culture and the culture around building commercial-military tech that surveils and values immigrant data. Then, I follow the story structure of the hero's journey to demonstrate how these companies are shifting the high-tech company culture toward a more military-commercial technology hybrid, showing some of the friction they experience as they attempt to embrace taboo data privacy violations.

First, let us establish the rules of the world as assumed by Silicon Valley and Western values around technology.

The Rules of the World are Established: Technology's Mythologies

When I teach my introductory course on information science and technologies, one of the most challenging concepts that my students struggle with is that technology is not a hero, savior, or inherently benign. We read books and articles on this topic; we conduct Google searches and notice the hierarchies of bias that Safiya Noble names algorithms of oppression. Nevertheless, at the end of the semester, the students feel compelled to add "but technology can be used for good," or "but the benefit outweighs the cost," unable to critique technology without defending it with a final word. What I see each semester is the strong pull of a cultural value. However, each week I ask students to suspend their disbelief and venture into critiquing technology as an "artifact" (Winner, 1986) that is not neutral. These are powerful societal myths that people are steeped in, and they are not as quickly named or cast with doubt as religious beliefs. However, these foundational beliefs are the rules of our world and the world of Silicon Valley and tech development.

Joseph Campbell said that mythologies are not human's search for meaning but our search to experience meaning. Mythology is the quest to experience living, "Myth helps you to put your mind in touch with this experience of being alive. It tells you what to experience . . ." (Campbell, 1988, p. 5). Story is also reproduced in our belief systems as a society through our mythologies. While we may not know or believe that we are engaging in myth in Western cultures, powerful mythological frameworks surround technology. Believing that technology is overwhelmingly good is a prevailing mythological and performative act. Just as Judith Butler taught us that gender is performing us (not that we perform gender) in everyday habits, we should also recognize the embedded, unseen power of stories and mythologies around and alongside technology.

Another prevailing mythology of Western culture is that technology is neutral and value-free. Social construction of technology theorists such as Langdon Winner (1986) argue that technology is a political and social construct, noting that it is embedded and created with its surrounding practices and dynamics of power. Technologies have political qualities. One recent example is how pulse oximetry devices were not accurately reading the blood oxygen level of people with darker skin tones (Lemoult, 2022). Technologies in everyday life are presented as value-free and designed without bias. That neutral cultural value represents technology *as* progress.

That forceful and robust pull that you might feel right now, to voice in defense: "yeah, but ultimately technology helps us and makes life better," or "what about all of the medical advances?" or, "what, do you want me to give up my phone?" is a part of our cultural norms and values. It is the force of the prevailing mythology that performs us. That is the pull of the mythology of technology in our everyday lives. The power in this myth is not that technology is a religion or value to which we strongly vocalize our allegiance, but that the myth goes unnamed and vigorously defended. Our values and stories around technologies are performative, as Butler noticed about gender. We do not perform those values; they perform us. Just as story performs us, these are the cultural values or the rules of the world in which this story takes place.

From here, let us dive deeper into the mythologies that construct Silicon Valley.

Silicon Valley Mythologies and Invisible Labor

The story of Silicon Valley, made into many successful movies and TV shows, tends to tell us about an innovative genius who had a breakthrough and developed a new technology that shifted our social structures for the progress of society. That story engages the arc of the hero's journey, the power of fiction, and braids in our religiously held value that technology equals futurity. I'm using the pronouns *our* and *us* here not to isolate non-American or non-Western readers but to acknowledge that I am just as much engaged in this story structure as many Americans and Westerners. This roving royal we includes citizens and technology consumers but only those immigrants who fulfill assimilation's potential.

Invisible Laborers

The prevailing myth in Silicon Valley is that one tech creator, usually a cisgender man, in his garage or dorm room or a nontraditional environment, built new digital platforms or specialized hardware that shifted our technological society and bettered our lives. I'll dive deeper into that myth soon. But let's first upend it. Authors in the history of science and technology studies and computing fields have demonstrated that computing is not created by one lone genius; rather, it takes many minds and bodies to complete the cycle of innovation, production, distribution, and consumption. Microsoft researcher Kate Crawford (2021) calls this web an atlas, or the atlas that makes up AI, visualizing the infrastructure layers it takes to build AI.

Venus Green's (2001) work on race, gender, and the social construction of operator switchboards in the telephone company reveals that marginalized and underrepresented people have long been involved in the assembly of technologies. Green braids infrastructure, technological development, corporate development, labor, and customer engagement to give a fully comprehensive history of how AT&T/Bell intentionally gendered the development of the operator switchboard and further dehumanized working conditions as required to employ more people of color. The author finds that telecommunication companies intentionally developed the telephone technology to minimize laborers such as the telephone operator, not because they were less efficient, but because they demanded better working conditions and better wages. Nevertheless, African American women had been telephone operators since the fifties (Green, 2001).

Mar Hicks (2017) pulls back the curtain of computing and reveals that women in Great Britain were the early computing workforce. As that country automated their work and boasted a technological revolution, tech labor became further gendered. Previous women laborers in that workforce were weeded out, their knowledge and skills lost, which left a noticeable gap that hindered the sophistication of technological innovation (Hicks, 2017). The story of the creation of the internet in the United States has traditionally been told as that of a military project; the work was assigned to military personnel. Joy Rankin's (2018) work on the people's history of computing demonstrates that the public was engaged in the internet's early development through programs in public universities across the country. As a result, the internet as we know it today was developed by many users' engagements, rather than solely the US military, Silicon Valley elite, or lone hackers (Rankin, 2018).

Infrastructure and information labor make up this bigger picture of invisible information labor. Many scholars describe the inner workings of our digital lives, the people behind granting us internet access. One example of this is YouTube. Algorithms cannot catch all the atrocities users upload, so information workers must screen red-flag content and determine if it should be removed. Sarah Roberts calls these workers *commercial content moderators,* "professional people paid to screen content uploaded to the internet's social media sites on behalf of the firms that solicit user participation" (2019, p. 1). They suffer debilitating mental and physical conditions as a result. Due to their invisible role and the nature of their work, these moderators rarely are revealed. Mary Gray and Siddharth Suri name the many people who work in businesses such as Facebook, Instagram, Twitter, and Amazon *ghost workers,* those information workers who work at the lowest rungs of these high-tech and multi-billion-dollar industries. They may be located all over the world, intentionally invisible but working as AI's backbone. "As builders create systems to transfer tasks from humans to machines, they surface new problems to solve through automation" (Gray & Suri, 2019, p. xxi). But that automation, in turn, leads to new problems ghost workers must troubleshoot.

There are many information laborers at different levels of these major tech conglomerates. Not all are counted as highly paid STEM employees or work under the famously luxurious working conditions of Google and Facebook campuses. Diverse Amazon employees who are known for their lack of representation in the valued STEM industries tend to work as delivery drivers for Amazon Prime and work in Amazon packaging facilities that are notorious for their exploitative conditions (Sainato, 2019; Crawford, 2021). Another example is the manufacturers of smartphones and the deadly labor conditions under which they are assembled. The most famous example is the iPhone manufacturer Foxconn, located in Longhua, China. About 1.3 million employees assemble iPhones in highly skilled and low-paid working conditions, which have led to notoriously high suicide rates at the plant (Merchant, 2017). The raw materials it takes to build a smartphone are also a lucrative and deadly business in the Democratic Republic of Congo, which supplies 60 percent of the world's cobalt. Apple, Microsoft, Tesla, and Google all have investments in the cheap mining of cobalt, which has led to child and adult deaths among the miners (BBC, 2019).

The foundations of computing have long employed people of color and marginalized people as integral components of data entry, technological assembly, and as citizen-consumers; but they have been engaged intentionally as invisible information labor and treated as disposable when that technology advanced in sophistication, their presence eclipsed as though they were never an integral part of a system.

Those information workers—the commercial content moderators, data scrapers, digital book scanners at Google, fiber optic engineers, iPhone production workers—play the foil to the Silicon Valley hero. They function as a nemesis of sorts to those heroes because they expose just *how* reliant information technology systems are on people of color and blue-collar labor. When visible, they act as a rival to the futurity of information technology because they also demonstrate how digital technologies are not simple, stand-alone items that make everyone's lives easier.

But the prevailing mythology exalts many heroes out of Silicon Valley. Tech moguls have framed their personal life stories, companies, and technology around the heroes' journey. Likewise, Palantir and Anduril use story structure to frame technological development around border technology. In Act I, the hero, the tech CEO, is called to an adventure: design technology that reaches beyond Silicon Valley's focus on consumers or entertainment. In Act II, the hero must overcome a great challenge by demonstrating the ability to "catch" immigrants with their technology. Finally, in Act III, they return to their world, changed: they return to Silicon Valley culture, now as hybrid defense companies, encouraging Silicon Valley companies themselves to expand beyond market consumerism.

THE SILICON VALLEY HERO'S QUEST

Silicon Valley mythology is founded on the arc of the hero's journey, framed by determination and meritocracy. The hero's journey was simply and most famously summed up by Campbell: "A hero ventures forth from the world of common day into a region of supernatural wonder: fabulous forces are there encountered, and a decisive victory is won: the hero comes back from this mysterious adventure with the power to bestow boons on his fellow man" (Campbell, 1968, p. 30). Sometimes this heroes' journey is called the "monomyth" because Campbell found that many storytelling practices worldwide followed this structure of mythology and storytelling. Fiction reigns in Silicon Valley lore, and the monomyth story structure is intentionally engaged for consumers and citizens to buy in.

The hero's story arc echoes that of the American dream. Hard work and determination drive a lone entrepreneur to invent a breakthrough technology, often dismissed or mocked by their peers. Eventually, this technology proves itself to contribute to the greater good. The entrepreneur is elevated from working out of their garage to investments by hedge-fund millionaires, breaking them through into the largess of tech company heads and improving society for good through new, efficient technology.

Silicon Valley itself recirculates and benefits from what Noble and Roberts (2019) call the "myths of a digital meritocracy." Silicon Valley mythology as of late holds dear to post-racialism, the ideology that race no longer matters, nor does it need to be discussed or acknowledged, especially in hiring practices. What follows is a firmly held belief that those tech moguls that made it did so through their hard work and gumption, also known as a *meritocracy*. The mythology of meritocracy and post-racialism in Silicon Valley has driven such trends as the mass gentrification of San Francisco, the East Bay, and Palo Alto and other peninsula suburbs. That same myth drives the lack of diversity in Silicon Valley white-collar jobs (Noble and Roberts, 2019). This myth advocates that those employed in the high-paying jobs of Silicon Valley earned their place through meritocracy. Their narratives of finding a high-paying job in tech are often cloaked in gender blindness and colorblindness: "Myths of a digital meritocracy premised on a technocratic post-racialism emerge key to perpetuating gender and racial exclusions" (Noble and Roberts, 2019, p. 4). Race and gender in Silicon Valley are acknowledged but assumed to be a subject of the past. The myth, then, is that Silicon Valley elites have earned their way to their success.

We can name many tech heroes' journeys off the top of our heads like they are Disney characters: Bill Gates wrote his first software program at thirteen (Encyclopaedia Brittanica, 2018), Steve Jobs dropped out of college and built the first Apple in a garage with Steve Wozniak (Hrnick, 2017). However, they are not always cast as pleasant or benign heroes. For example, the fictional portrayal of Mark Zuckerberg in *The Social Network* portrays him as an undermining business partner who builds Facebook out of the ingenuity of coding prowess and backstabbing, a portrayal, it turns out, that is not as far from the truth as he once claimed (Carlson, 2010).

The techno-hacker nerd is often portrayed as an embodiment of hyperrational masculinity, and that hyperrationality is associated with genius and intellect. For example, the movie portrayal of Zuckerberg as blunt, out of touch with society, and socially inept is established as congruent with the genius behind IT development, rather than problematic (Sweeney, 2020; Eglash, 2002). That gendered rationality is positioned as out of touch with reality but makes up for that with genius. Tech moguls' hyperrationality maps onto white masculinity.

The nerd trope has been established as a type of pejorative but represents someone with unmitigated access to science and technology. The white male nerd archetype has been noted by scholars as a signifier guarded by whiteness and gender (Eglash, 2002, p. 60). Black, Indigenous, and other people of color have been intentionally removed from the association with science and technology but have nevertheless been present in tech history, albeit invisible.

The hero begins with nothing, downtrodden or sidelined in some way. Then they use their gumption to innovate something new. People often dismiss them based on their foreignness, and they struggle to prove their value, but they later triumph when their technology is found valuable by users and investors. Any association with marginalization or overcoming barriers is considered a valuable trait in reflection on their story. One example of this is the exalted Christian Gheorghe, a Romanian immigrant who was once a limo driver and became the founder of the business analytics company Tidemark (McBride, 2013). Gheorghe is esteemed by his venture capitalist colleagues because he applies himself to hard work and becomes successful, despite his background. However, that ideology of meritocracy is assumed and applied to all successful entrepreneurs and employees in Silicon Valley. "It's an impressive tale that encapsulates the way Silicon Valley likes to think of itself: a pure meritocracy; a place where talent rises to the top regardless of social class, educational pedigree, nationality or anything else" (McBride, 2013).

However, a 2013 analysis that dives deeper into that myth reveals that most Silicon Valley companies were led by entrepreneurs with prestigious degrees from Stanford, Harvard, or MIT, personal connections, and investments (McBride, 2013). While the underdog, immigrant-made-millionaire stories are celebrated and normalized by Silicon Valley mythology, they are the exception and not the rule. Woven into interviews by tech giants is the hero's journey, framed by meritocracy: "Silicon Valley has this way of finding greatness and supporting it. . . . It values meritocracy more than any place else'" (Joseph Ansanelli of Greylock, as quoted in McBride, 2013).

The Borderland Hero

The above framework is the mythology our own heroes/antiheroes of this story enter: Anduril's Palmer Luckey and Palantir's Peter Thiel and Alex Karp.

Many players in the Silicon Valley surveillance tech collaborative could be our point of focus. Some other possibilities would be Jeff Bezos of Amazon, Sean Mullin of BI2, or leaders of the many companies chapter 3 focused on. But I consider Palantir and Anduril to be the leading companies in this burgeoning industry. The Silicon Valley hero/antihero merits a closer look as we dive deeper into the mythologies framing borderland tech. The dislikable tech nerd archetype is a long-established trope with power. Early tech industry founders presented the façade that their innovations were framed by liberal and neoliberal values of social equality by creating open-access products, information for all, and public access to digital networks, advocating that those technologies would further democratic values (Weigel, 2020). However, new leaders of Silicon Valley have emerged that are invested in alt-right ideologies, often with connections to alt-right news sources such as Breitbart or investments in alt-right leaders such as President Donald Trump (Weigel, 2020).

Founder of Anduril Palmer Luckey is a hardware guy. He dropped out of California State University, Long Beach, in typical Silicon Valley tech self-made fashion. He became a millionaire by building the virtual reality (VR) goggles design breakthrough for the Oculus Rift (Manthrope, 2016). Luckey describes himself in the true self-made, tinkering-in-the-garage fashion of tech mogul hero history: "I'm a self-taught engineer, a hacker, a maker, an electronics enthusiast. I'm from Long Beach, California" (Purchese, 2013). Luckey's father was a car salesman, and he was homeschooled by his stay-at-home mom, taking community college courses in his teens (Purchese, 2013). As a kid, Luckey was looking for a solid pair of affordable VR goggles for video games and was not able to find them, but he found a niche and need in the gaming market. Building tech in his garage and perfecting the design of 3-D and VR goggles, Luckey founded Oculus VR in April 2012 through a Kickstarter campaign with several big tech endorsements and investors, raising 2.4 million (Purchese, 2013). By age twenty, he was a millionaire.

Tinkering and serendipity are a common theme in Palmer's bio, according to a write-up on the then twenty-year-old Luckey: "He tinkered, modified and invented while the experts he'd hired did what he didn't know how to do—develop the all-important Oculus Rift software developer kit" (Purchese, 2013). We can hear the echoes of the long-told mythology around Silicon Valley heroes and their humble beginnings: "In the space of four short years, Palmer Luckey has gone from being a regular Joe in his parents' garage, building VR head-mounted displays, to the head of a 30-person team and the face of virtual reality gaming"

(Purchese, 2013). Oculus VR was acquired by Facebook in 2014. Luckey is now a known alt-right supporter and donor to Trump's 2020 presidential campaign. In 2017 Palmer left Facebook and his association with Oculus, or he was possibly fired in the controversy of his openly supporting the Trump campaign. In 2017, Palmer founded Anduril Industries, and the company began a pilot program in 2018 along the US-Mexico border in a bid for the virtual border wall funds.

In the above outlined call to action, the Silicon Valley tech hero merges Silicon Valley innovation with a defense technology mission by leaving Silicon Valley. However, the purpose or vision behind the quest is demonstrated in much of Luckey's language to describe his company. In the Silicon Valley–defense industry merger, goodness is usually projected as a patriotic duty. For example, in 2019 he held a ribbon-cutting ceremony in Irvine, California, for his new technology defense company. He had wanted to use his replica Lord of the Rings sword, named Anduril and carried by Aragorn in J.R.R. Tolkien's famous novels, for the ribbon-cutting ceremony, but he didn't have time to sharpen it (Dean, 2019). Nevertheless, he drew on the Lord of the Rings mythos to denote the importance of this event when announcing his intentions:

> "Anduril," he said, leaning into the long Elvish vowels, "means Flame of the West. And I think that's what we're trying to be. We're trying to be a company that represents not just the best technology that Western democracy has to offer, but also the best ethics, the best of democracy, the best of values that we all hold dear." (quoted in Dean, 2019)

The rules of the world for Silicon Valley are that meritocracy and personal drive are the deciding factors in personal success. Politics tended to lean toward socially liberal but fiscally libertarian, with the overriding belief that technology development does more harm than good. Anduril and Palantir came into this world with alt-right affinities. Still, they propagated the myth that *their* tech could be a hero when taking up the call to action—to merge commercial and military defense technologies.

PALANTÍRI: FANDOM THROUGH THE DATA BODY STATE

In Act II, the hero leaves the commercial product development of Silicon Valley and ventures further into applying those commercial technologies into defense in the borderlands (broadly and digitally defined). In this case, the result is the creation of Palantir and Anduril.

For these two companies, accepting the call to adventure meant pivoting away from Silicon Valley consumer tech and accepting the challenge to build widespread surveillance technology. Western cultural story structures (and global one) engage the employees, consumers, and citizens indirectly participating as surveillance informants.

Palantir was established in part with CIA funding and has had its foot in the consumer and government contracting worlds since its inception (Alden, 2016). Established in 2004 by Peter Thiel, an early investor was the CIA's venture capital arm, In-Q-Tel (Alden, 2016). In 2016, Palantir was said to be struggling with their retention of employees and retaining customers, who were said to pay around $1 million per month for services. However, from Coca-Cola to Kimberly-Clark, many customers found that Palantir did not provide returns on their investments, and they ended their patronage (Alden, 2016).

In Silicon Valley's merger with borderland defense technologies, story provides context for citizenship and technology design. Although a religious belief system underlies the world where this story exists in technology's ability to do good over evil, the story line is the framework on which these companies build their product. Story is a world that employees enter as they help construct surveillance products. Moreover, algorithms function to retell the normative story in many ways, one without the context of structural inequalities. Moira Weigel sums it up like this:

> In place of interpretation, data analytics substitutes correlations that it treats simply as given. To a machine learning algorithm that has been trained on data sets that include zip codes and rates of defaulting on mortgage payments, for instance, it does not matter why mortgagees in a given zip code may have been more likely to default in the past. Nor will the algorithm that recommends rejecting a loan application necessarily explain that the zip code was the deciding factor. . . . Algorithms take the histories of oppression embedded in training data and project them into the future, via predictions that powerful institutions then act on. (Weigel, 2020)

The power of the story in these big data sets is that it repeats itself incessantly. It draws on previously established stories—tropes, stereotypes, and systemic oppressions—and reaffirms them with more data.

These supporters of alt-right politics emerge as a visible divergence from previous Silicon Valley tech CEOs, who presented themselves as aligning with socially liberal politics. In contrast, their companies' actions and investments apparently align otherwise (Weigel, 2020). Perhaps one of the best examples of this shift in Silicon Valley and the American zeal for technological innovation is Anduril Industries. To tell that story, we need to go all the way back to another white guy tinkering in his garage.

We can consider Palantir casting the border and military defense style surveillance as one large "Con" (as in Comic Con), wherein they are cosplaying the good guys of the *Lord of the Rings*, à la projected augmented reality. The playing board, or gamescape, is the United States as a digital border, and their employees are engaged in the story line of fandom.

Palantir's name draws on Tolkien's classic fantasy world, where "like the crystal ball, Palantir can not only track data but use data to predict behaviour, raising red flags among civil liberty watch dogs" (Green, 2019). As many readers will know, the Lord of the Rings is a classic hero's journey story about the hobbit, Frodo Baggins, who must go on an adventure and leave the comfort and safety of his world, the Shire. Frodo, who has inherited the Ring of Power, the one ring to rule them all, from his uncle Bilbo Baggins, must take the ring to be destroyed in the fire of Mount Doom. Frodo travels with the Fellowship of the Ring, which consists of hobbits, humans, an elf, and a dwarf. *The Lord of the Rings* manifests in storytelling in movies, video games, and merchandise as well as the books. Along with Palantir, Thiel previously referenced the Lord of the Rings through his companies Valar Ventures, founded in 2010, and Mithril Capital, which he cofounded in 2012.

The Lord of the Rings lore frames Palantir's mission and vision. That story casts Palantir CEOs and employees as the heroes in a big adventure. Their logo is a Tolkien palantir viewed as an orb or else an open book, reflecting Palantir's goal of human-computer symbiosis (Simonite, 2014) (fig. 4). In Tolkien's book, a palantir, also known as the "seeing stone," is an orb that can be used for communication at far distances, to see one's enemies' secrets, and to see events in the past or future ("Palantíri," n.d.). *Palantíri*, the plural form, means "far-seeing" in the elvish language Quenya ("Quenya," n.d.). In the film adaptation of *Lord of the Rings*, the palantir is often a harbinger of corruption and misinformation. For example, the human king Denethor used a palantir to see a black fleet of ships sailing and believed they were coming to attack his kingdom, misunderstanding that it was Aragorn leading those ships. In *Fellowship of the Ring*, the white wizard Saruman uses a palantir to communicate with the villain Sauron to build armies for

FIGURE 4. Palantir Pavilion at the World Economic Forum, Davos, Switzerland, featuring the company's logo. Photo by Cory Doctorow, 2017, via Flickr.

Mordor. In *The Return of the King*, Peregrin Took touches a palantir and is attacked by Sauron. Sauron also uses the stone to weaken Aragorn as they confront Arwen's other in the final battle at Minas Tirith, where Sauron uses a palantir to show Aragorn that his beloved Arwen's life is fading ("Palantíri," n.d.). The elves created the *palantíri* to communicate and predict their enemies' plans; the examples in the books and films mostly demonstrate that they are destined to corrupt and be used by villains. Perhaps this conveys a cynical self-awareness by Palantir about the nature of their company's work.

Anduril also echoes this good/hero and bad/villain binary. Anduril is named after the powerful sword in *Lord of the Rings*, the sword forged from the shards of Narsil and used by Aragorn, known as the Flame of the West. This literary reference positions undocumented borderland people as the villains of *Lord of the Rings*—which can include any villain from Sméagol/Gollum to orcs to Sauron—but these are villains who can be defeated with a mythical tool. Data body milieu, then, is a state that can code immigrants and refugees as illegal and villainous, designating the tech company itself as the hero of a story arc. In the data body milieu, the capture of these villains is centered on the capture and housing of their data.

Palantir employees role-play heroes of the Lord of the Rings story, and the surveilled targets are the villains. This involves not only misguided namesakes but also fandom as a tool for immigrant information surveillance. Palantir employees are welcomed into consumer and defense projects through an already built-in Lord of the Rings fandom base.

While Lord of the Rings is the central fandom focus, other employees tend to use a general idea of heroism in accepting the company mantle. In a 2012 *Forbes* interview, one employee discussed their investment in the vision of Palantir, "Indeed, Palantir sees itself as a collective of unlikely heroes—think X-Men with glasses—who are out to defeat evildoers wherever they may lurk" (Quora contributor, 2012). As discussed in chapter 3, Palantir's Gotham database again casts any city using the database as the crime-ridden city that Batman battles in DC comics media. The Palantir employee can identify with multiple hero figures in media fandom.

Mel Stanfill, a scholar in texts and technology, notes that fandom is an essential tool for media industries and is "increasingly seen as a resource to be managed whenever possible" (2019, p. 3,). In the business world, a fan "refers to interactive audience, where 'audience' means the act of membership in an audience" (p. 5). Recently, media companies have viewed fans as more interactive than solely consumers, and fans have become valued central and mainstream media engagement forces (p. 6). As a result, media companies manage fandom for their own benefit, essentially "domesticating" fandom (Stanfill, 2019).

Stanfill notes that media and sports-consuming fans are *assumed* to be white cisgender males as an unmarked category marker: "The whiteness of fans is not neutral or inevitable, but rather the product of power relations" (2019, p. 25). Fans, particularly gaming, speculative fiction, and anime or other convention fans, are assumed and understood to be white men. However, for Palantir, what is essential is an unspoken tone that casts the Palantir employee, whether it is their identity or not, as the assumed white male fan. Research has demonstrated that Tolkien fans have an invested unawareness of whiteness and insist that race is "not happening" in fandom (Stanfill, 2019). Speculative media fans are assumed to be white audiences, evidenced by the overwhelmingly white *Lord of the Rings* heroes—from the elves to the hobbits. Whiteness is cast by both the heroes and the assumed fandom of speculative fiction.

This is not to say that all Palantir's workplace is white cisgender men. However, by being cast within the fandom of speculative fiction culture—specifically Lord of the Rings—there is an assumed or implied whiteness. But also, as Stanfill puts it, "whiteness shapes the discourse of the fan" (2019, p. 56).

Palantir has cast their company as a Lord of the Rings cosplay in name and mission. Their offices are named The Shire, Grey Havens, Rivendell, and Gondor, all locations throughout the *Lord of the Rings* books. Palantir is "mapped" through the Lord of the Rings landscape and engages the employees as active players to defeat Sauron's greater mission. So compelling is this cosplay that we see a sort of fandom built around the Palantir cosplay. Through Glassdoor, one Palantir employee titles their post "Save the shire!" and states, "This is a terrific place to work if you get excited by working on tough, exciting, and important problems with incredibly smart people" (Glassdoor contributor, 2014). "Save the Shire" is an internal Palantir mantra (Green, 2019), and Palantir's yearly conference is called Hobbitcon.

Stanfill's term *coercive positivity* explains why Palantir builds their company around fandom. For fans, coercive positivity is "the demand that fans produce love and only love" (Stanfill, 2019, p. 155). In the employer rating website Glassdoor, Palantir doesn't have exceptionally high and gloating employee recommendations. Of those 370 reviews, 78 percent would "Recommend to a Friend," 86 percent "Approve of CEO," and 70 percent have a "Positive Business Outlook." In qualitative ratings, those more glowing recommendations reflect the fandom surrounding the company's work mission, structure, and leadership. Among the cons listed on Glassdoor, employees regularly comment on the Palantir "flat structure" of the work organization culture. In 2012 an employee said of his work at Palantir: "For an organization with over 600 people—and growing its team rapidly—Palantir is still remarkably flat. You have your co-founders, then a couple of directors followed by a handful of leads—everybody else is essentially an engineer" (Quora contributor, 2012). The flat structure is often identified as the modus operandi for Palantir, attributed to the hero role-playing. While the company has hierarchy through CEOs and managers, the perceived equality of all employees lends itself to the Silicon Valley myth merged with fandom mythology—that anyone in any role within Palantir can "save the Shire."

The mythological layout for Palantir helps cast a fictitious augmented reality where employees are always in the cosplay world of their employer. Fiction in Palantir has a role to play, and it provides a stronger pull than nonfiction facts, "Fiction seems to be more effective at changing beliefs than nonfiction, which is designed to persuade through argument and evidence" (Gottschall, 2013, p. 150).

Villains and Bad Guys

Frequently the labels used to justify technological development in the borderlands include "drug runners," "coyotes," "criminals," and "bad guys," along with conjured images of immigration en masse; the scale of many people or the hostile border desert also considered villainous for their uncontrollability.

A long-established criminalizing trope against Latinxs crossing the border is held to threaten the US, Western, and white normative way of life. The Latinx threat (Chavez, 2008) is an early narrative that militarized borderlands and kept many Latinxs on the undocumented side of citizenship while reinforcing nativism. "Bad guys" on the border has been long applied to Mexican immigrants. As reviewed in chapter 2, legislation directed at Mexican immigrants began to shift toward hostility in the early 1900s, with growing popular sentiment for quotas limiting Mexicans more than European immigrants: "Anti-Mexican rhetoric invariably focused on allegations of ignorance, filth, indolence, and criminality" (Ngai, 2004, p. 53). Throughout borderland policy, rhetoric may shift depending on what serves the state best. So the question becomes, what function does this contemporary emergence of the bad guy on the border serve for a state–Silicon Valley merger? Justifying these emerging technologies relies on bad hombres but also on a threat of masses on a larger scale, as Coolfire Core demonstrates in their mock-ups for proposed digital wall technology:

> Concerns over border security have come to the forefront in recent years, prompting both government officials and the general public to demand a more effective system for monitoring border traffic. Now, with a caravan of over 8,000 people from Central America rapidly approaching the border between the United States and Mexico, finding a sustainable solution has become more urgent than ever.
>
> While thousands of migrants are still in transit, several hundred have already attempted to cross into the U.S., resulting in dangerous border clashes—clashes which in the future might be avoided through the use of smart technology. (Coolfire Solutions, 2018)

A tweet by Palmer Luckey on July 4, 2020, about the developing borderland technology demonstrates the villain/hero binary that structures this emerging industry: "Immigration policy is a hotly debated issue with a broad spectrum of rational opinions, but every reasonable person (including every politician I have ever spoken with on both sides of aisle) can agree that we need to know what happens on our borders so we can stop the bad guys" (Luckey, 2020).

Technological design is fueled by a long-rooted fear around Latinx immigrant influx and an always present bad guy. AI promises to bring the Latinx immigrant in the borderland from undocumented to documentable data through correlating data. The networking of these old and new technologies converts data into information. In the *Lord of the Rings* movie, the villains were cast as people of color. Sauron and most of the orcs are played by Samoan actors, whereas most heroes are white. In Palantir and Anduril, the villains, or targets, are usually immigrants and communities of color. Story structure and the hero's journey are heavily embedded into borderland tech development, dividing the players into the good-guy tech creators and employees, border patrol, ICE, and tech users/consumers versus the bad guys (figs. 5 and 6). The bad guys, always insinuated but never fully named, are Latinx immigrants in the borderlands. While the borderland technology is applied to all on the border, the coded language of drug runners, coyotes, and cartel Latinx men is the specter that shapes this particular time period and supports enhancement of the emerging collaboration between Silicon Valley and DHS.

Coolfire Core, a company that builds commercial and military technologies including apps, demonstrates the prototype of the gaze that falls on immigrants as data bodies on the border. They describe the emerging data border as the alternative to the physical wall:

> A "digital wall" would rely on a network of deployed connected sensors and cameras to monitor border traffic, prevent unauthorized entry, and enable border security operations to make well-informed decisions in real-time.
>
> With data provided by relatively low-cost remote sensors, and integrated into an effective situational awareness platform, field units could achieve the operational vision necessary to identify and prioritize potential dangers, responding appropriately to all relevant threats. (Coolfire Solutions, 2018)

The Coolfire Core graphic "How a Digital Wall Works" on the same page demonstrates it by visualizing the command and control center at the far end of the image, a security outpost, a ground sensor, and a stick figure at the border walking past a ground sensor. The stick figure silhouette is the always unnamed presence in proposals for technological design.

FIGURE 5. Surveillance towers loom over the US-Mexico border wall. Crosses memorialize those who have died along the border. Photo by Tomas Castelazo, 2006, via Wikimedia Commons.

FIGURE 6. Customs and Border Protection (CBP) officers view the border through images taken by unmanned aerial vehicles. Photo by Gerald Nino, CBP, 2006, via Wikimedia Commons.

For undocumented people, being undocumented is already considered a criminal act. An ICE spokesperson said that the agency's National Fugitive Operations Program teams "target and arrest criminal aliens and other individuals who violate our nation's immigration laws for the safety and security of our communities" (ICE, n.d.). The spokesperson added that 90 percent of those arrested by ICE "had either a criminal conviction(s), pending criminal charge(s), were an ICE fugitive, or illegally reentered the country after previously being removed." The ICE statement also said such individuals are "subject to immigration arrest, detention and—if found removable by final order—removal from the United States" (Piven, 2019). However, in 2017 documents obtained by Mijente demonstrated that ICE was targeting, arresting, and deporting anyone suspected of being a noncitizen with the National Fugitive Operations Program (Mijente, n.d.). The labels of criminal, aliens, or fugitives could be cast as a wide net to cover anyone undocumented.

What's in a Name

Databases are critical in current modes of deportation. For example, databases are used to store biometric data, store general identifying information, and apply AI to make predictions and conclusions about that information.

For example, a stand-alone database at a state's department of motor vehicles may hold identifying information, forms, images, and biometric data. However, once that system is connected to another system—for example, a department of corrections database that holds biometric information from everyone who has ever been incarcerated—you have a network of systems working together to inform one another. That is where we are today with immigrant surveillance and deportation. So many databases, networked to one another, determine where an undocumented person is, fill in the information gaps around that person, and lead to their incarceration or deportation.

In 2019, *Vice*'s Motherboard obtained a Palantir user manual for Palantir's database Gotham, the database that law enforcement agencies use to pull up networked data on a subject. The user guide demonstrated that police could bring up data about a person, starting with very little information, such as just a name associated with a license plate. From there, the database can reveal email addresses, phone numbers, addresses, bank accounts, social security numbers, personal relationships, physical descriptors, and networked relationships (such as the data of family, friends, and businesses). All data pieces that describe a subject are called objects; objects are broken down into entities, events, and documents (Haskins, 2019).

The way the Gotham database works is that the user searches several tools to organize data to make up a profile of an individual. For example, the Easy Search Helper and starts with a license plate field or a personal name, and from there brings up networked information such as emails, bank account numbers, physical features, and personal financial information (Haskins, 2019). The law enforcement officer can also use the Histogram Helper, a chart that connects objects to each other in a web, essentially networking information to one search subject. "The Histogram Helper calculates and displays the frequency of individual Objects, Properties, and Link types selected in the current application" (Palantir, p. 1). The Heatmap Helper also visually demonstrates the concentration of objects on a map, like the Object Explorer, which visualizes data points in different charts.

One of the noticeable practices of Palantir, Amazon, and other tech companies like Anduril is that they rarely name undocumented immigrants within their design demonstrations, but rather reference a silhouette of a villain that has long eluded the US government's grasp. The use of the term *object* to profile a person in the United States is telling: a name is no longer needed to build a profile on an individual. An individual's profile is built up of data points through networked data such as a license plate number. It replaces that human person with a data body. Bad guys, criminals, cartel, sex traffickers, and drug traffickers are among the labels employed when referring to undocumented people as data targets. One example of this was Donald Trump's allusion to *bad hombre* in his speeches leading up to the 2016 presidential elections (CNN, 2016). Another example was Palmer Luckey's July 4, 2020, tweet discussing Anduril's borderland technology.

In other databases, a name may be everything. For example, Blacks and Latinxs made up 87 percent of the CalGang Criminal Intelligence System database, which 92 percent of agencies use across California. A 2016 audit of the database found that most of those listed were babies under one year old, associated as gang members based on "some combination of zip codes and racially coded names" (Benjamin, 2019, p. 6). CalGang (also known as GangNet) "allows user agencies to collect, store, and share intelligence information about individuals suspected—but not necessarily convicted—of being involved in criminal activity. . . ." (California State Auditor, 2016, p. 2). CalGang has also failed to protect individuals' data rights to privacy (p. 23). The way information privacy is invaded is that individuals' information is collected based on the suspicion that they are affiliated with a gang, without actual evidence of criminal activity having been collected or documented (p. 28). In this instance, the ethnic-sounding names became justification for those individuals becoming networked data subjects.

Names and networked identifiers that build an individual's data body, or data profile, are "cultural coding" embedded into technical coding (Benjamin, 2019, p. 9). For undocumented immigrants, Blacks and Latinxs especially, their name is the launching platform for these law enforcement databases to build their data bodies. Often for undocumented immigrants that go unnamed or are perceived as a threat because of their lack of data (in the form of social security numbers or drivers' licenses), those data bodies are built around every other identifier that can shape them into a legible, surveilled subject.

THE HERO RETURNS, CHANGED

Silicon Valley heroes experience multiple shifts after they accept their quest to transition from commercial market design into defense. I want to talk about three of these changes. First, commercial technologies change *based on* immigrant surveillance data, shifting Silicon Valley and the military-industrial complex into a Military-Industrial Startup Complex. Second, border patrol and users of this hardware and software are *changed,* rewiring their story reflexes through practicing on this new reality. And finally, the Silicon Valley hero discovers themself metamorphized and must contend with that change and adjust the plot.

The Military-Industrial Startup Complex Metamorphosis

While government and industry use the Latinx threat as a platform for funding, building, remodeling, and testing these new technologies, perhaps even more alarming is the more significant push to merge the military complex of technology design with consumer, civilian technology. In 2019, AI became all the rage, especially for its potential in borderlands. Anduril's Luckey critiques the collaboration between tech business and the Chinese government, though the model nods to the growing collaboration between U.S. defense and technology consumable products: "In China . . . there's no difference between their civil sector and military sector. It's all one thing. That's why when you help Chinese companies work on artificial intelligence, you are almost directly helping the Chinese military work on artificial intelligence. A lot of people in Silicon Valley don't really understand this . . ." (London, 2019).

New technologies that have already been introduced into the consumer market—such as VR goggles, LiDAR technology for self-driving cars, and drones, are perfected on the surveilled person—in this case, the Latinx immigrant.

Story as Dress Rehearsal: Practicing Reality on the Border

The heroes of the borderlands are also altered intentionally through virtual reality (VR) and augmented reality (AR). A reminder of the story outlined above: fiction and story are more potent than nonfiction or factual evidence. Stories, fiction, and feelings move humans to act. Psychologists Melanie Green and Timothy Brock argue that "entering fictional worlds 'radically alters the way information is processed. . . . [W]hen we read nonfiction, we read with our shields up. We are critical and skeptical. However, when we are absorbed in a story, we drop our intellectual guard. We are moved emotionally, and this seems to leave us defenseless'" (quoted in Gottschall, 2013, pp. 151–152).

Fictional experiences, such as movies, television shows, books, and video games, serve a purpose beyond entertainment: they light up parts of our brains as if we are experiencing it in real life. We practice stories to prepare for real life (Gottschall, 2013). The means by which the brain practices for real life by way of fiction is via "mirror neurons." "Practice is important. . . . [A]ccording to evolutionary thinkers such as Brian Boyd, Steven Pinker, and Michelle Scalise Sugiyama, story is where people to go to practice the key skills of human social life" (p. 57). Mirror neurons are simulations in our head, but interestingly they are inaccessible to the conscious mind (Gottschall, 2013). That doesn't mean fiction is not powerful. In this simulator model, the human brain cannot determine the difference between fiction and nonfiction; the brain still processes that rehearsal as real life. Gottschall calls this effect "brains on fiction": "When we experience fiction, our neurons are firing as much as they would if we were actually faced with that real life scenario. . . . [W]e equip ourselves for real life by absorbing fictional game plans" (p. 64). "The constant firing of our neurons in response to fictional stimuli strengthens and refines the neural pathways that lead to skillful navigation of life's problems. . . . This is because human life, especially social life, is intensely complicated and the stakes are high" (p. 67).

Fiction in its old form is changing into new forms. Video games are one example of how people are experiencing deeply immersive storytelling: ". . . most video games are organized around the familiar grammar of problem structure and poetic justice. . . . [T]he games are usually lurid but heroic violence narratives. Such games don't take their players out of the story; they immerse them in a fantasy world where they get to *be* the rock-jawed hero of an action film" (Gottschall, 2013, p. 182).

While the form has changed, the function has remained in mirror neurons *practicing reality*. For example, the function of storytelling on the US-Mexico border through border patrol comes in their use of VR and AR to practice for hostile unknown situations and challenging terrain (USA Today Network, 2018).

Virtual reality emerged as a tangible technology in the 1990s, when video games promised the possibility of VR headsets; however, nothing developed in VR until the past decade. In 2009 the Wii device was introduced, which was considered the first AR and VR experience. By 2010 PlayStation and Microsoft had introduced their own VR games. In 2012 Oculus Rift created a Kickstarter project to launch a realistic VR gaming set and raised over $2.4 million, with their first demo launching in 2013 (Stein, 2019).

In 2013, Google Glass was introduced to the public, which, while failing to become popular, opened the door to many other VR glasses from other manufacturers. The following year is considered to be the year that VR broke through with advancement, and it was adapted and advanced by many tech companies; Facebook bought Oculus for $2 billion. In 2015 and 2016, VR made another more advanced jump with sensor technology through HoloLens, Google's AR Tango phones, the iPhone X's Face ID camera, and the PlayStation Move with controllers (Stein, 2019). Finally, 2016 was when many VR devices were introduced, including the Oculus Rift, HTC Vive, PlayStation VR, Google Daydream, and Microsoft's VR headsets. That year, Oculus founder Palmer Luckey was on the cover of *Time* magazine for the future as VR (Stein, 2019).

In 2017, consumers experienced AR through games like Pokemon Go or the use of Ikea's app to place virtual furniture onto physical, real-world spaces. VR and AR had split to accomplish different, though related, tasks: "But there was one key difference now: AR's killer trick involves spatial awareness of the world, something that early VR lacked. That spatial camera computer vision magic would play into a lot more areas of tech, including face scanning, autonomous vehicles and drones, and security cameras" (Stein, 2019). Then, in 2018, Luckey began pursuing another field altogether: defense and border technology.

Ready, Player One?

As VR and AR improved over the years through consumers' data, the US Border Patrol also became a first-person player by engaging these emerging technologies in the borderlands and around the Unite States. Since Luckey entered defense technology, the media has noticed, "there's a futuristic, gamer feel to Anduril's products, which, going back to Luckey's background with Oculus, bring a kind of virtual reality simulation experience to border surveillance" (Ghaffary, 2020). We see the echoes of the quantified self, the human brain and body being hackable through popular neuroscience and biotech quantification by coupling VR and AR with border patrol. VR and AR are welcomed as tools that can finally conquer the terrain of the US-Mexico border or enhance border patrol officers' minds and bodies to conquer complex scenarios through practice.

In 2016, the US Border Patrol invited press members to experience virtual reality in the borderlands. Fernanda Santos with the *New York Times* reported her experience. Each player experienced a simulated engagement that followed a story line based on the player's response. According to the report, the instructor who trains border patrol agents wanted agents to see their interactions and responses to various scenarios. VR was justified as a means to train agents not to react immediately and know when force or shooting at an undocumented person is necessary (Santos, 2016). So, again, we come to justifying the use of advanced technology for altruistic reasons. In the demonstration in 2016, Santos described how border patrol was using VR to essentially stop themselves from shooting everyone through realistic first-person player border scenarios.

Border patrol has used VR goggles to scan the border using the infrastructure of previously installed surveillance towers (Shankland, 2020). DHS's Science and Technology Directorate collaborates with border patrol and federal law enforcement to provide night vision technologies or VR goggles along the border. These goggles are described as resembling the first-person video game *Call of Duty*, in which undocumented people on borderlands are cast as enemies in modern warfare. The user of the VR goggles (a border patrol agent) is positioned as the United States or its allies, the good guys or heroes, in any given war from history

(Shankland, 2020). Video games that take place along the US-Mexico border present what media scholar Juan Llamas-Rodriguez (2021) demonstrates as borrowing from colonial tropes and ideologies that reproduce race and violence along the border's developing technological infrastructure, "by creating playable narratives that invoke the untamable frontier and position racialized subjects as Other" (para 2). Llamas-Rodriguez reveals how tunnel vision within these borderland video games demonstrate how border infrastructure, in contemporary VR and AR technologies developed by Silicon Valley, shapes race and colonial violence.

VR and AR were built around games and story lines that profit off the tropes of Black and brown people. In political discourse, first-person shooter video games such as *Grand Theft Auto* threaten white male players through the Black and brown secondary characters (Leonard, 2009). David Leonard names the culture of gaming, particularly VR first-person games, as oriented in and built around white supremacy. The first-person player is imagined as white (Leonard, 2009). In the first decade of the twenty-first century, politicians introduced legislation around violent video games. Those political debates focused on the "youth" and protecting the player's innocence against entering "into virtual ghetto spaces, places defined by hypersexuality, violence, criminality, and a disregard for the rule of law" (p. 253). However, there was a lack of public discourse on the white imaginings of Black and brown people and spaces envisioned by the story line of those video games, which notoriously have Black and Latinx characters and neighborhoods as hypervisible centers.

The archetype of gamer had traditionally been reserved for nerdy white males. Some gamers defended that title when people of color and white women became visible in the gaming culture. One example of this was the online attack #GamerGate on visible critics of the limitations and exclusions around gaming. Gaming theorist Megan Condis argues that gamers are performing in the opposite direction of Judith Butler's theorizing around gender performativity. In Butler's analysis, gender is performative in that it is something we do every day, not the essence of who we are. Condis (2018), working off this, notices how gamers apply gaming into all systems that they encounter: "Thus, the field of gendered and sexualized (and racialized and classed) identities becomes one of many playing fields, a space to be inhabited strategically. For gamers, gender identity is a contest that can be won" (p. 7). Within gaming, the ideologies and function of gender and race became embodied online through repetition and performance. Masculinity is performed through gaming with privilege. In game design, Condis advocates that "we need to pay attention to how they are made, as well as by whom and for whom" (p. 10).

In 2017, the military began employing AR in an inverse version of VR's general use, with the Augmented Reality Sandtable (ARES). The Army Research Lab describes the ARES as "a traditional sand table which is augmented with low-cost, COTS [commercial, off-the-shelf] technologies to offer novel, multi-modal interactions and visualizations" (STTC Research, 2017). This sand table gives a 'battlespace visualization' to build a multidimensional terrain visualization. In 2019, border patrol publicly demonstrated their use of the ARES, alongside night vision technology, to conduct sign cutting and tracking training in an AR setting (Department of Homeland Security, 2019). "ARES serves as a low-cost solution that gives USBP the capability to create terrain for a specific area of interest dynamically, visualize lines of sight and overlay different types of maps to create a 3-D hologram" (DHS, 2019).

In the use of AR for digitally documenting the border, we see what Shoshana Magnet calls the physical border—bemoaned by border agents and tech companies alike for the challenges it presents to navigate and map—transformed into an inverse gamescape. The gamescape *is* the imagined and virtual landscape created by video games. However, for the AR Sandtable and border patrol, the borderland becomes the gamescape, practiced on the ARES to control wild terrain. We may think of the use of AR as an attempt to dominate wild terrain, finally. "Imaging landscape" is *how* space is constructed and imagined through video games. ARES is a US colonial imagined landscape that attempts to control a terrain through imagination and practice: "The ARES interactive digital sandtable uses augmented reality to create a 3D map of any given terrain. A projector displays a topographical map of the desired environment on top of sand in a large sandbox, to visualize features of the terrain like trails, valleys and hilltops" (GCN Staff, 2019).

AR and VR are training tools that help border patrol *as trackers* visualize the terrain along with their targets. Under DHS, sign cutting and tracking training merges multiple technologies for borderland training. For example, one study within DHS reported that two groups of newly hired border patrol agents were given either the old-school, legacy sign cutting and tracking instruction, or training on the DHS Science & Technology Directorate's (S&T) newly developed computer-based training for two hours. Through this research, DHS found "both groups then went out into the field for their practical exercises to demonstrate their skills in 'tracking boxes' that instructors used to assess their tracking performance," and that the agents who were "trained using S&T's new Sign Cutting and Tracking Training solution performed 63 percent better than agents who received the legacy classroom instruction" (Kimery, 2019).

Those trackers that used AR/VR techniques were considered augmented beyond those who had solely trained outside. AR and VR are ways that border patrol can practice their reality: this technology uses storytelling to train the brain and body to react more quickly in real-life borderlands. This quantitative self changes with the use of storytelling: VR and AR are designed on the foundational story that immigrants in the borderlands are drug smugglers, child traffickers, and other types of villains, and the VR and AR are types of weapons that can augment their instinctual response time to stop the villains. VR/AR has physically changed the hero through these overarching tropes around immigrants.

The Hero Becomes the Supervillain

In a third demonstration of the Act III story arc, or how the hero returns to their world a changed person, we revisit when Palantir as a company has come to terms with its villainous traits. By 2020 the company had become known for its work with the federal government, and most notoriously, their work with ICE. Palantir demonstrates the growing pains of moving from the story's heroes to national villains as protests increasingly emerge within tech companies and consumers.

In a 2020 interview with HBO, *Axios* reporter Mike Allen asked Palantir CEO Alex Karp about his work. Allen asked, "Was there ever a time that you wished that you had not done work with ICE?" Karp responded, "Absolutely, completely." Karp then went into detail about his employees leaving over ICE, his very liberal family expressing anger about his work with ICE, and the subsequent activism and protests against that work: "I've asked myself, if I was younger in college, would I be protesting me?" (Allen, 2020). Karp says about the US government's use of Palantir for intel and clandestine services, "Our product is used on occasion to kill people" (Allen, 2020).

Palantir decided to move their headquarters from Palo Alto to Denver. Karp expressed that Palantir had moved beyond Silicon Valley's persona as he filed paperwork for Palantir to go public in August 2020: "Software projects with our nation's defense and intelligence agencies, whose missions are to keep us safe, have become controversial, while companies built on advertising dollars are commonplace. For many consumers, internet companies, our thoughts and inclinations, behaviors and browsing habits, are the product for sale. The slogans and marketing of many of the Valley's most prominent technology firms attempt to obscure this simple fact" (Levy, 2020).

In Palantir's good-bye letter to Silicon Valley, Karp realigns Palantir with the hero model through the merger of the defense and commercial sectors. However, the company recognizes that it cannot maintain a flawless hero persona: "We embrace the complexity that comes from working in areas where the stakes are often very high and the choices may be imperfect" (Levy, 2020). Like Frodo and Bilbo on the elven ship, heroes must leave home, as they have changed and don't belong there anymore.

PART TWO

Reimagined Techno-Futures

Pero Queríamos Norte

How do we respond to the billions of bytes of data mined to surveil immigrants? How do we imagine a different state than the one we are already so enmeshed in? How do we account for technologies that are pitched to consumers as the next flashy Black Friday gift, yet were originally designed around immigrants in borderlands?

The immersion and state of unknowing around many surveillance technologies make this state of data body milieu seem unsurmountable at times.

In Part II, I propose short parables in response to the ubiquitous state of the data body milieu. In opposition to the archetype of manual laborers, Latinx immigrants are technologically savvy, using digital tools for everyday life and as an addendum to citizenship. I *imagine* future borderlands and technologies with my community in Riverside County, a highly surveilled borderland among wineries and suburbs. I used imagined techno-futures, a Black feminist methodological approach to oppressive technologies, to speak alternatives to the emerging data border state.

Gomez Peña noted the friction of being viewed as an outsider to online cultures, while also being present and immersed with technology. Imagining Latinx immigrant communities' techno-futures with people from my community who are targets of the companies that this book discusses, emphasizes what Gomez Peña long ago observed:

> The role for us, then, was to assume, once again, the unpleasant but necessary role of cultural invaders, techno-pirates, and coyotes (smugglers). . . .

> What we want is to "politicize" the debate; to "brownify" virtual space; to "spanglishize" the Net; to "infect" the lingua franca to exchange a different sort of information—mythical, poetical, political, performative, imagistic; and on top of that to find grassroots applications to new technologies and hopefully to do all this with humor and intelligence (2001, p. 197).

As a response to the surveillance dragnet, Part II offers vignettes of the nuances in Latinx immigrant borderland experiences. This section attempts to serve as an alternative response to a large and powerful underlying story around information technology. This approach engages Black feminist socio-techno and futurist theory and critical information science scholars as a call for change in the ways we view, regulate, and engage information technologies.

When we spoke about the state of surveillance in borderlands and around the country, occasionally my interviewees would say "*pero queríamos norte*" or "*pero querías norte*" which would usually conclude or explain a conversation about the exploitation of immigrant experiences in the United States. It is a meme, used in some popular corridos, or can be a rallying call to push through and get to work, despite the difficult conditions. This directly translates to "but we wanted the north" or "but you wanted the north," said in different scenarios. At times it applies when people are not looking forward to work but reminding themselves of why they made the journey. This phrase gives agency back to those who immigrated by acknowledging that they knew the xenophobic and exploitive conditions in which they would work but chose the circumstances of undocumented life in the United States. They knew what the stakes were, and they made the journey anyway: "*pero queríamos norte*" takes agency away from anti-immigrant circumstances, border patrol, and exploitive coyotes and gives it back to Latinx immigrants who travel north. It's a saying that embraces the many frictions of immigrant life. *Pero queríamos norte* is a way to bookend and give the final word on data borders back to undocumented people and Latinx immigrant communities.

THE ANTIDOTE TO STORY

The antidote to a story is a parable. For a story to be a parable, the reader must have been changed after they read it. According to popular theologian Pete Rollins, parables are different from stories: "Parables do not substitute sense for nonsense, or order for disorder. Rather, they point beyond these distinctions, inviting us to engage in a mode of reflection that has less to do with fixing meaning than rendering meaning fluid and affective" (2016, p. xi). What does a parable do? A parable is not a parable unless it changes the listener, it "cannot be heard without being heeded" (p. xii). Rollins says:

> This truth can be spoken only by those who live it and heard only by those who heed it. These timeless incantations have gone by many names over the millennia, but one such name is 'parable.' In the parable, truth is not expressed via some detached logical discourse that would be employed to educate us, but rather it emanates from the creation of a lyrical *dis*-course that inspires and transforms us- a *dis*-course being that form of (mis) communication that sends us spinning off course onto a new course. (2016, p. x)

When thinking about parables, perhaps one of the most resonant parables of our time comes to mind: Octavia Butler's *Earthseed* series. Butler wrote *Parable of the Sower* in 1993. In 2020, the plot of *Parable of the Sower* became hauntingly predictive, as it includes an authoritarian president, peak environmental destruction due to climate change, social unrest from racism, a pandemic, and more familiar crises that hit home hard. Butler's main character is a teenage Black girl named Lauren Oya Olamina, who is spreading a new set of ideas called "Earthseed." At the heart of Earthseed is change: "All that you touch you change / All that you change changes you / The only lasting truth is change / God is Change" (Butler, 2012, p. 2).

Butler's parable has become a foundational toolset for adrienne maree brown's emergent strategy, a community organizing principle that many activists of color utilize in various movements for social justice. Author and activist brown, along with musician Toshi Reagon, have dedicated multiple art forms to Butler's parables, including music, an opera, and a podcast. For brown and Reagon, the main takeaway is change. Reagon and brown promote adaptability, change, and positive obsession as a response to crisis, marginalization, and as an avenue of liberation. And that change, they note, must be done individually through both personal and systemic change.

Brown states, "In movement we orient around what needs to be changed as an external directive. Those people that are 'bad,' who are ignorant, regressive, capitalist . . . they need to change, and we need to change that. . . . [T]his line troubles me . . . how would this thing that I touch change me?" (Reagon & brown, 2020). Reagon and brown advocate for personal change and societal change: "Its only what we're practicing, what we're willing to touch, that can change. We have to be willing to be in that transmutation together. . . . [Y]ou have to be willing to get in and get dirty, and be changing" (Reagon & brown, 2020).

A parable not only changes our opinions or perceptions but changes our actions. To be a parable, the mainstream way of thinking, or story, must change. An antidote to story structure would erode the deeply held beliefs around technology, citizenship, and state formation. But unlike story, parables can't tell the audience exactly *what* to believe, they can only radically change that audience; Jesuit priest Gregory Boyle says, "Parables don't tell you what to do and they have no didactic endings" (2021, p. xiii). In *Parables of AI,* Singh and Guzmán (2022) use parable to ". . . build a *community of storytellers and listeners living with data and artificial intelligence (AI)-based systems in the majority world*" (p. 2).

I approach border futures through imaginings as parables and antidotes to the storytelling that happens from and around oppressive information technologies. Techno-futures through imagination is one method as a response to data borders, big data, technological policing, and AI in our borderlands. This section aligns with what Catherine D'Ignazio and Lauren F. Klein have labeled *counterdata initiatives* (2020, p. 35), a resistance to the majority narrative of big data through small data and techno-future imaginings.

BLACK FEMINIST THOUGHT AND IMAGINATION AS A METHOD

When I first outlined this book, I intended to include a chapter demonstrating immigrants' experiences with surveillance technologies. Multiple journalists have written about these experiences, and they importantly place immigrants at the forefront of the story (Bedoya, 2020; Funk, 2019; Grinspan, 2019). However, as I come to the end of this book, I want to focus on those emancipatory approaches that Black feminist writers have advocated through the method of imagining. This part of the book embraces more liberatory futurisms of tech in conversation with undocumented people. My method prioritizes individual Latinx immigrant narratives of their borderland stories, as well as reimagining borders, space, and techno-futures. This approach resists creating a technology as a direct response to a societal issue by putting this conversation in a bigger context of inequities and liberatory imagining. In opposition and contrast to big data, this method values one Latinx undocumented person's insight and experience into technology with the weight of millions of bytes of data.

In reflecting on Afrofuturist and Chicanofuturist approaches, Ruha Benjamin asks us to reimagine technology for liberatory purposes:

"This work has a lot to teach us about reimagining the default settings—codes and environments—that we have inherited from prior regimes of racial control, and how we can appropriate and reimagine science and technology for liberatory ends" (2019, p. 195). An emancipatory approach to technology critiques the oppressive designs of tech and entails "a commitment to coupling our critique with creative alternatives that bring to life liberating and joyful ways of living in an organizing world" (p. 197). Safiya Noble does this by ending her work with "The Imagine Engine," a different type of search engine that would visualize the indexed web in a transparent interface that considers raced, gendered, and sexist bias. Noble notes the importance of imagining alternate futures for the technology that promulgates systemic oppression: "such imaginings are helpful in an effort to denaturalize and reconceptualize how information could be provided to the public vis-á-vis the search engine" (2018, pp. 180–181).

In her seminal text *Emergent Strategy*, adrienne maree brown builds an approach to social change that embraces imagined futures: "We hone our skills of naming and analyzing crisis. I learned in school how to deconstruct—but how do we move beyond our beautiful deconstruction? Who teaches us to reconstruct? How do we cultivate the muscle of radical imagination needed to dream together beyond fear?" (2018, p. 45).

Popular relationship psychologist Esther Perel discusses why storytelling through imagination is so important to marginalized communities. For Perel, this was evident in storytelling practices through her family after surviving the Holocaust:

> The storytelling is part of what created continuity. What creates continuity in all communities that are uprooted, dismantled, broken. Everybody, there's nothing unique in here, this happens to be my personal Jewish holocaust story, but every community, worldwide: storytelling binds you to people, binds you to the past, binds you to the transmission. Freedom in confinement comes through your imagination. . . . When you are physically incapable of leaving, the only place that you can leave through is your creativity, your mind, your imagination . . . stories, imagination, that goes hand in hand. (Perel, 2021, quoted in Brené Brown podcast, 2021).

In confining situations where marginalization has shaped family structure and individual circumstances, storytelling and imagination must be the methods of freedom.

This book is about the data borders most US Americans reside in and the data bodies from which Silicon Valley profits. There are new technologies invested in and applied every day at the intersection of commercial and military technology collaborations, and they are built off the visions, missions, and imagined innovations that distribute normative imagined identities. These technologies reinstate normative imagined identities of citizenship, race, gender, sexuality, and class. They put people back into their "imagined" normative places through software and hardware: AI, data mining, algorithms, drones, virtual reality goggles, and more. But brown (2018) advises that her method of emergent strategy must use alternate imagined futures to erode these normalizing dystopias: "I often feel I am trapped inside someone else's imagination, and I must engage my own imagination in order to break free" (p. 15).

Influenced by Benjamin, Noble, and brown's methods of imagination as a counter to systemic oppression, the next chapter first tells Latinx immigrants' physical borderland experience through their experience of crossing into the United States. I argue that when immigrants speak their physical border-crossing stories, we put bodies back into the data body, and the data borders we all reside in are made tangible. Their stories said out loud and shared give us an alternative to the quantified immigrant's data body. Second, we imagine alternative borders, data bodies, and technologies by imagining new borderlands and technologies. I interviewed people from my own community who came to our region (Riverside County) without papers and are of various citizenship statuses now, ranging from naturalized to still undocumented. Whereas traditional academic practices might gather data and analyze the results, I want to maintain that the work of techo imaginings and borderland futurity is not from first gathering the data and then analyzing the results but happens when undocumented people share their stories in their communities.

CHAPTER 6

First-Person Parables

Imagining Borderlands and Technologies

Almost a century ago, my grandpa grew up in the borderlands of New Mexico, Texas, and Mexico. At some point, while walking in a desert somewhere between Las Cruces and Ciudad Juárez, he heard a woman's voice. "Turn around," she echoed out of thin air. When he turned, he was facing a rattlesnake in attack position. He grabbed a large rock, aimed for the snake's head, and crushed it before it struck. Every time he told us this story, he would end with a lesson, "and that's why I always take a walking stick when I'm hiking." Growing up, I often heard stories of migration from Mexico to the United States and back. The stories usually came with a life lesson, and miracles and mysteries were woven among the travelers' tales. These border tales were odyssean in my childhood mind: The stakes were high, the travels were dangerous, and the hero was always trying to find home. As I've spoken to family and friends over the years, I've noticed that many people who came undocumented across the US-Mexico border experienced their own odyssey, complete with the highest stakes imaginable, life-and-death scenarios, villains and heroes, miracles that require the suspension of disbelief, and long journeys that seek home.

SETTING

Border patrols are often seen driving around cities in Riverside County, California. The Newton-Azrak Station is best known for its Temecula

Border Patrol Checkpoint. The original checkpoint was established in May 1924, in the Temecula Valley, as a remote site known only as Temecula, California. The Newton-Azrak Station, responsible for an operational area covering 3,800 square miles, maintains the checkpoint on Interstate 15, a secondary checkpoint on Highway 79 at Oak Grove, and several checkpoints throughout the area (US Customs and Border Protection, 2020). This valley is known for using various surveilling technologies, it's within 100 miles of the Tijuana border and 100 miles from the ocean, as of time this is written is within jurisdiction for electronics to be searched and seized and for social media to be used for purposes of surveillance for detention and deportation (Theodore L. Newton, Jr., 2020). The Murrieta Police Department uses the Amazon Ring surveillance camera system, and job ads for ICE agents local to Murrieta suggest that the employee must know how to search social media, RELX, and Thomson Reuters (see box 1).

Murrieta became nationally known when anti-immigrant activists blocked a bus of incoming, already detained, migrants they didn't want to allow into their city (Farberov & AP, 2011). More recently, cities like Murrieta, Wildomar, Lake Elsinore, and Temecula adopted anti-immigrant legislation in 2010–11, leading up to the 2014 bus blocking. In 2010, with lobbying from the local anti-immigrant Tea Party movement, cities in Riverside and Orange County adopted anti-immigrant policies mirroring those that had been recently passed in Arizona, such as symbolically embracing Arizona's controversial SB-1070 law that legalized racial profiling in policing (Cuevas, 2010). All Temecula Valley city councils voted to require the use of E-Verify, a DHS and Social Security Administration program that checks the legal status of new workers, until Governor Jerry Brown signed the Employment Acceleration Act of 2011, which among other points forbids cities to use E-Verify (McAllister, 2011).

Box 1 is an example of the extent to which ICE agents are entrenched in the merger of Silicon Valley educational tools. Note how this posting reads like an IT professional's job description. This ICE senior analyst job description, posted for my hometown and located in McLean, Virginia, described the job as data based: "Our team will be conducting open-source analysis to identify potential threats at the nexus between illegal immigration, national security and public safety."

The people I interviewed live under intense surveilling circumstances in and around the Temescal Valley. They have had regular interactions with border patrol and ICE, beyond their border crossing experience. For the interviews, I prioritized data privacy through several avenues:

Box 1

Re-created job description for an ICE Senior Analyst position, listed in Murrieta, California

ICE SENIOR ANALYST HSEP9

- Our team will be conducting open-source analysis to identify potential threats at the nexus between illegal immigration, national security and public safety.
- Experience analyzing publicly available social media information. Experience using Thomson Reuters's CLEAR and/or LexisNexis Accurint systems.
- At least 10 years of continuous experience in performing primarily social media and open source information, research, targeting and analysis
- Avantus is looking for a Senior Analyst to manage a small team to locate illegal immigrations who pose a danger to national security or public safety.
- Master's or other advanced degree
- DHS suitability

I have university IRB approval for ethical research practices, kept data private, and used pseudonyms for written responses. I received grant funds from various sources and used funds to pay people for their time. In addition, I only interviewed people who know and trust me personally, a method I view as antithetical to the anonymous and consentless practices of big data and algorithmic data gathering.

One counterdata and counternarrative movement that has emerged is the "Feminist Data Manifest-No," which resists the implied neutrality of data that removes data making from power. In resistance to the common practice of using data as a method to marginalize, the authors state, "Our refusals and commitments together demand that data be acknowledged as at once an interpretation and in need of interpretation. Data can be a check-in, a story, an experience or set of experiences, and a resource to begin and continue dialogue. It can—and should always—resist reduction. Data is a thing, a process, and a relationship we make and put to use. *We can make it and use it differently*" (Cifor et al., 2019).

In the spirit of the "Feminist Data Manifest–No," there were frequent discussions during my research period around the ways in which Latinx immigrants cross and evade technology surveillance with their own innovative creations, discussions that I did not record, document, nor plan to discuss here. All I will say is that there is ingenuity, technology creation, and innovation in the geographic US-Mexico border in response to heavy surveillance.

I summarize the interviewees' border-crossing experiences and included their responses to my questions about reimagining the border. All names have been changed for anonymity. Some interview collaborators chose their pseudonym and some left me to decide on one for them. Because these interviews focus on my community, everyone I interviewed is Mexican; however, Riverside County is diverse with immigrants those from the Philippines, El Salvador, and Guatemala.

First I describe the biographical and border-crossing experience of each interviewee. I try to keep the responses I translated about imagining techno-futures as close to direct quotations as I can. Some sentences that stood out in Spanish I kept verbatim in Spanish. I tried to keep the story intact and translate directly and in the first person. from the "I" perspective. The power of oral histories isn't solely in the data that arises from these conversations. The agency happens in real time when people tell their story out loud from their perspectives and experiences. Putting the body back into the data body requires agency for undocumented people to have a safe context to speak their experiences of border crossing, technologies that surveil, and reimagine different techno- and border futures.

LUZ, TERESA, CRISTINA, MARI: *UN PINCEL DE RAPUNZEL*

I interviewed a mother in her late fifties, Luz, and her two daughters, Teresa (age twenty-four) and Cristina (twenty), on their experiences crossing through the Tijuana border in 2001, months after the September 11 attacks. Her third daughter Mari (eighteen), who was born a citizen in the United States, was also present and contributed to our conversation. For many that came before September 11, 2001, permanent residency and citizenship were more available. However, after that date, US citizenship became much more difficult to obtain. Luz wanted to come on a plane, but didn't want to fly. She tried to get a visa but couldn't obtain one. Her husband had a visa and would travel back and forth from the United States to Mexico, but eventually he stayed and decided to cross them.

Like many people in Lake Elsinore and surrounding areas, Luz came from Guanajuato, Mexico. Luz crossed with her three young children and told me her story, starting with how she didn't want to come: "*La verdad es que yo no quería venirme a los Estados Unidos.*" Her border crossing was terrifying, because she was separated from her three young children. Luz's husband was sending money to bring her two daughters (three years old and six months) and her one-year-old son. She believed they would cross together, but when they got to Tijuana, the coyote (smuggler) told her they would take the kids first and then her. One driver told her they would follow the car her kids were in to arrive at a stranger's home in another part of Tijuana.

When they arrived, they were in a room with strangers and none of her kids. In the Tijuana house, Luz was crippled with guilt and fear that she had handed over her small children to strangers. She felt as though she had lost her soul while waiting at the Tijuana house. She looked out the window constantly, but the car that was going to bring her girls never arrived.

"*Es una cosa que no so le deseo en nadie,*" she told me.

Coyotes at the house said they wouldn't be able to take her across. They took her to another house, far in Tijuana. The house didn't have a door on the bathroom, there were mattresses all around, and a lot of people were sitting around. The coyotes wouldn't feed them unless they gave money. An old woman told Luz that if they her for money, she shouldn't show it to them, because they would take it. Luz wasn't hungry anyway. She stayed like that for two days. The women in the room would take turns watching over each other while they slept.

The next night, coyotes arrived and said if they wanted to, they could move them to another house. It was late at night. A few started asking her why she wanted to leave, that they would take her out on dates. Luz and two other women broke out of the house and ran. She told the woman not to cry because they would be heard. They ran to a main street, and she thanked God because a bus picked them up. They went to the Tijuana airport and called her husband. After three days, she found out the kids were with her husband in California. She asked him to bring her kids back; she didn't want to do this again. Her husband told her that he would find more honest and safer coyotes.

She crossed over the border, trying to use an ID with a picture that didn't look like her. La Migra took her out of the car. They let her out and told her not to cross again. Later, she tried to cross again. "I saw border patrol and put it in God's hands." She requested to God that if this would be for good, to help her, and if it would be for a worse future, to send her back. This time she was able to pass through, and a girl in the car told her: "We did it, we made it."

Teresa, the oldest child, remembered having her siblings with her. Before they were separated, her mom told her to take care of them. Teresa remembers her siblings crying. The coyotes told her they needed to stay quiet or they would be left behind. She told the women driving the car that they were hungry, but they didn't get any food. Teresa remembers other children in the car with them and moving to different cars. Eventually, after transferring into many cars, they stopped, and when she got out of the car and got out, she saw her dad. "I remember when we got to my dad, I couldn't stop hugging him, and everyone was laughing because when I got to my dad, I couldn't let him go. I didn't see my mom for a very long time and asked when we were going to see her."

They moved to Lake Elsinore with family and grew up there. Border patrol and ICE would frequent businesses or places where immigrant workers stand to be picked up for work. When their father went out and La Migra would raid the city, their father would have to hide. They would randomly raid apartments, stores, or pick-up spots for day laborers. The men who were looking for work were always the most targeted. Their father often barely escaped ICE raids.

At the time of these interviews, Teresa, Daniel, and Cristina remained undocumented. Subsequently, the California DREAM Act was passed in 2011, but only Teresa and Daniel were able to receive protections under the DACA executive order. When Cristina wanted to apply for DACA, Donald Trump became president and DACA was revoked. Daniel, Cristina, and Mari attend University of California schools, and Teresa is waiting to enroll as a dental hygienist. DACA protections were not reinstated under President Biden until October 31, 2022.

How would you imagine the borders if you could make it anything you wanted?

CRISTINA: I feel like they should humanize the whole process. They have been trying tactics like prevention through deterrence, where they will leave certain parts of the border open, and immigrants will come. They leave those parts open because it's a tough trail, men are going to die. They depend on immigrants to pass the word that it's a tough passage, many people will die. They underestimate how people are desperate . . . many of these people are desperate to cross, they're going to keep trying. . . . To humanize the whole process and create a better understanding of where people are coming from. Why they are choosing to leave.

In the training of border patrol agents, they'll do unnecessary things to make the process more cruel, like they'll take water away in the desert. They take advantage of immigrants' positions because of the rhetoric spread around the country. Just educating the border agents, keeping in mind what kind of people are put in those positions. You'll learn a lot of the border patrol are Hispanic, why would they put themselves in the position to torture other Hispanics who are struggling? They are desperate to fit in with white American values. Educating, doing as much as you can to create a more understanding narrative.

More open borders, I feel. If these people do so much to get to America, they will do so much to succeed. People talk about drug cartels, but most of these people are not carrying anything. More open borders will benefit America overall.

How would you imagine Lake Elsinore if you could make it anything you wanted?

LUZ: For the university to be closer where my kids attend; a science park where they teach kids what they could be interested in, somewhere where they could mentor kids.

TERESA: I would bring back the tutoring house that we had during our childhood. When my siblings and I have our late-night conversations, we talk about how much that changed our childhood.

How would you imagine or reimagine border patrol?

LUZ: How do I feel about them? If I see them, I run the other way. I wish they would take into consideration the situation of each immigrant. If you're born here, you have it all, it's not the same experience [if you are undocumented], many people don't have what they have here in other countries. I wish they were more conscious of their positions, to make better decisions. With people, at the end of the day, we're all men and women, we're all brothers and sisters.

MARI: I wish they were more educated on why immigrants come here. Like, [in the United States] the facts that we learn about other countries are very vague, they [US citizens] don't understand how it is in other countries. If I talk to anybody about immigrants, they don't know much about how the process works. No one I talk to knows about it unless they have a family member in their house. People don't know, so they assume they know.

Or even how they say immigrants are stealing jobs, we know it's jobs Americans don't want to do.

TERESA: Because they don't know, they think people come here as a choice, like they want to go through. They just want them to get their papers.

CRISTINA: I don't believe border patrol should exist.

How would you imagine citizenship?

LUZ: I just wish they were more conscious of our kids, the opportunity to give the kids the opportunity to grow and to study. Our countries don't have that opportunity. To have social security, to build their dreams and their careers, or more inclusively, projects . . .

I would go be with my mom when she had an illness. I would go see my siblings.

TERESA: My parents applied for citizenship in 2001. When we tracked the wait list, it was eighteen more years. It's been twenty years since my mom has seen her mom. My dad had to go on Skype to see his mom's funeral.

If you could create any technology in the world, what would you create?

MARI: Teleportation technology, because it would be easier for my mom to see my family (in Mexico) that she hasn't seen in twenty years. . . .

LUZ: "*A mi me gustaría que existiría un pincel de Rapunzel.*" I would use Rapunzel's paintbrush so I could paint anything and go through the painting. I would go through to my mom's house.

TERESA: Something to sort through the citizenship process faster and efficiently. Why does it take so long? Are they going person by person? A way to sort through the process faster. They have technology for everything, but they can't use algorithms for applications. It's difficult to believe, but who's working on it?

Anything I didn't ask that you would like to say?

LUZ: I wish that the kids at the border waiting, that the border patrol would be more straightforward if they're going to cross or not. And that immigration was more compassionate of the governments of other countries that were looking for the opportunity to work. That they could create jobs around the border to give the people jobs [in detention centers], they're able to do those jobs.

TERESA: There are a lot of immigrants who don't want their children to apply to DACA; you must give so much information, a lot of the parents are nervous because you're giving them everything. Where we entered, every house we've resided in. Every two years when I do my renewal, I have to do my thumbprint and pay for my thumbprint. I have to give them my parents' names, siblings, it's everything. God forbid something changes; they know where to find you and your family. When it was new, you didn't know if it was legit, if there were going to be repercussions.

I didn't know I had any limits until I was in high school. Our parents don't talk about it because you're not supposed to talk about this with anyone, it's a secret that you're undocumented. My teachers wanted me to do dual enrollment with a college class, but the teacher told us we had to bring in our social security number the next day. I was afraid they would know because I wasn't coming to class. When you're sixteen, everyone gets cars. At that time you needed a social to drive. It was one more thing I couldn't do. Or my friends would get jobs. With DACA it kept me going, but for my sister it was different. At least with DACA I had a social and it was some type of protection. The fees for school are higher, you have to pay out of state, you're considered out of state, even when there is a waiver. They're afraid to use a waiver [for undocumented students]; there is so much fear to do anything.

CRISTINA: When everyone was doing the FAFSA in college, I was excited. Then they said, "bring your Social Security." But no one mentioned the Dream (DACA) at all. The school counselors didn't mention DACA at all. I had to find someone else undocumented, but I was afraid to out myself.

In elementary I didn't know that because I was born over there, I didn't have papers. But I didn't know I didn't have papers.

I still allow myself to dream big, but then I realize there are limits. . . . I wanted to be an FBI agent, and someone wanted to introduce me to someone to get an internship. But I don't have a social.

DACA was open for a while, I got my prints in, but then it closed. I had applied for an internship with a California senator, but there are things that I want to do, and I can't do them. It feels like me and my brother are stuck here, we can't do anything. . . . We don't want to disappoint my mom and all of her sacrifices, but there is little we can do here. Should we move somewhere where they would give us residency? It feels like we're telling my parents that their sacrifices weren't enough if we leave.

People don't care at all; they don't know how undocumented people live . . . even immigrants in general. One night my mom asked if she had done us wrong, because of everything we have to face on our own. I am making the most of it.

LUZ: Yo les digo que no pierdan la esperanza, que tengan fe. . . . Not to think they haven't been doing anything—they've sacrificed a lot, they've shown dedication, God is with them and will support them and only he knows how he will help them later. I have faith that God will help them. I've asked God. In my past I told you that I didn't want to come. But if it was for good, let me cross, if not send me back home. It was a sign that he wanted me to be here. I have faith there will be a way.

JUANA: YOU LOOK UP AND JUST SEE STARS, CLEAR, AND IT WAS BEAUTIFUL

Juana is a Mexican woman who was born in Guanajuato, Mexico. She came to the United States in 1989 when she was nine. She received her permanent residency when she was in high school, at nineteen or twenty years old. Her border crossing story involved traveling with her brother and mother by bus from Irapuato to Tijuana, and then going to a hotel and staying for a few days. Her mother contacted a coyote, and they took another bus to the border up *el cerro* (a hill):

> There were a lot of people there. The coyote got there and told us the instructions, that we would run, whatever happened, it didn't matter. We had to run, even if someone in my group got caught, the rest had to run. We were hiding in a big group, when we saw immigration coming, we got scared, so everyone was running everywhere. I saw my brother run with an uncle, but when we saw they got caught, we turned around and broke the rule, we said we were going to stay with him.

At that point they went into detention.

> We were put in a van, and they took us into detention in the United States. They took us to a big containment room. I remember it was cold, I'll never forget about that because I was freezing. A lot of people like us in there, many of them crying, many of them sleeping on the floor. Everyone was depressed. I remember the officers checking on us, how were we doing? Asking us questions, tak-

ing our fingerprints. Since I was little, the toilet situation was really embarrassing to me because everyone had to go in the same room, and I was shy.

They released us to TJ, we were released a day later, it was daytime. So we go back to the hotel and try the same thing. Some people have to pay all over again, but this coyote was my stepdad's friend, and they made a deal, so we didn't pay again.

This time we made it. We crossed at night. It was cold as hell, I don't remember what month. I remember the big lights, big poles with bright lights at night. You could hear the quads. They had big trucks, we were hiding somewhere, you would see them pass you, and then you run and hide in another spot. At that time, I remember we had to run. The second time we moved we crawled like little soldiers. You pause, you must pass inspections. There were times we waited a long time, sitting down in the bushes. It was dark, cold, you look up and just see stars, clear, and it was beautiful.

The coyote knows what time they're going to pass, everything. They spend a lot of time studying every move. Right now, it's harder because they have a lot of technology, but back in the day it was easier.

There is somewhere you can buy cars close to the border. Once they said the driver was coming for us, it was a relief, no more helicopters.

I was mad the whole time. I was thinking about how we were going to have to learn English, make new friends, go to a new school. I was crying because it felt terrible leaving the rest of my family. *Abuelitos*, *tíos*, *primos*, and neighbors. Leaving the house I grew up in was very hard for me.

They picked us up by the freeway, and they took us to a house, and there we waited for another ride for days, three days in a house.

Her family moved straight into apartments in Lake Elsinore that were targets known to border patrol. Border patrol and ICE were a regular presence, using scare tactics of randomly showing up at homes and workplaces:

I remember being in the play area of the apartments where we lived, and I would see a lot of men running. One time my mom grabbed us when they were running around the apartments. They were asking us questions. That was every week, they would park outside the apartment complex. It was Sundays or during the week when most men came from work. I would see them in the streets, close to the schools when they were going to pick up their kids.

They stopped people based on the color of their skin, how they dressed. They would target the guys dressed like cowboys or dressed in construction boots.

How would you imagine the borders if you could make it anything you wanted?

Safe. I understand people are coming illegally, but it's sad that some die on the way here. It should be safe. I wouldn't mind more patrolling if they can help people, like sometimes the coyotes leave kids behind, it would be nice if it was safer.

How would you imagine or reimagine citizenship to look?

I agree with some of the government's rules, which is like you must be here for a certain amount of time. I would make it not as severe, giving people the chance to be here and show us how they are as citizens.

If you could create any tech in the world, what would you create?

Something that cleans the air, anything that has to do with nature, like the air and water. I would want something so we can get rid of plastics. I like taking care of the environment.

As far as technology, it would be good if there were robots that help people on the border like if someone was dying, the robot would be programmed to call for help, but this wouldn't be used against the people crossing. Like no jail time, just send them back. Also, there should be a kind of technology that helps find people who have passed away [on the borders], like if they are buried, the machine would be able to see bones underneath. This way families would have some peace and are able to say goodbye to their loved ones.

Anything I didn't ask that you would like to say?

What makes me sad is the people that were undocumented and went into the military services, and then just got deported. In La Línea [the border], they were just deported. They served, and then they were just dumped out there. I think if they defended the country, we owe them more.

OSCAR: I WOULD CREATE AN APP THAT LETS PEOPLE START OVER WITH A FRESH START

I was born in 1984 in Irapuato, Guanajuato, Mexico. I was four years old when I crossed with my mom and my sister. My sister and my mom were together. My stepdad's brother was our coyote. I remember the border patrol had their headlights on us on their quads, there was border patrol on horses. They surrounded us and we gave up, and the next thing I remember was that they had us surrounded. The detention center was cold, and I asked my mom if they were going to kill us, because the whole time they were talking to us the border patrol had his hand on his gun, so I was terrified for my sister and my mom. I think it was two attempts, but we got to the US when I was around five. My stepdad and my mom rented a house in Lake Elsinore.

Border patrol would keep an eye on the Mexican markets, where a lot of men would sit and wait for someone to pick them up to work. The border patrol would raid these poor men. They didn't mess with kids or women; it was mostly the men. My mom would tell me to keep an eye out for them and run from them.

How would you imagine the borders if you could make it anything you wanted?

I think the border should stay closed but that they would have sensors to see if someone is not crossing with drugs or anything else, let them into the other side and stay in the United States, give them a try to apply and stay.

How would you imagine here, Lake Elsinore, if you could make it anything you wanted?

I think they should be equal about cleaning the city. By that I mean getting rid of the drugs, because Lake Elsinore is pretty, but these druggies make it not safe for kids. There are syringes on the playgrounds. Adding sidewalks and more streetlights.

How would you imagine or reimagine border patrol?

I wish that most of the border patrol weren't Hispanic, it's such a disgrace attacking your own people when they're just trying to create a life in the US, just like they have. First- and second-generation Hispanics think they're above illegals, whether it's Mexican, Salvadorean, Guatemalan . . . it's just embarrassing.

If you could create any technology in the world, what would you create?

I would create an app that lets people start over with a fresh start. Something that could give immigrants the option of not having a record, clearing their record. Like for people who have been deported two or three times . . . like people who change their names, giving people a new start.

AZTECA: *CUANDO MIRO ARRIBA ESTÁ LLENO DE ESTRELLAS*

If it was up to Azteca, she wouldn't have come to the US. She was told by many people that life would be hard without papers and without a car. Azteca left Mexico to get away from her abusive partner, her first child's father and because there are more opportunities for people than in Mexico. School is free. If people are hungry, they're given food. She wants more opportunities for her children.

She first tried to cross at the border checkpoint in a car, but when they arrived, she decided it was too dangerous and returned. When she tried to cross again at Tijuana she was caught by La Migra. She told the agents that she was trying to cross for more money, more opportunities, and they sent her back. Later, a relative asked her if she wanted to cross through to the United States, and she said yes because her son was already there.

During her third try, she crossed at night through the mountains at night in Mexicali. She found a coyote who would take her but was hesitant because she was pregnant, it would be dangerous, and it would take all day. Azteca told them that she wanted to see her son who was already in the United States, so they started walking. Halfway through the day, the border patrol started following them. The helicopters, the dogs, and the *motos* were out. They continued to walk. She feared snakes because it was hot. Then it was dark. When they came to the hill that was cleared out with a tower with a bright light on at night that swiveled around, they knew that it was an infrared camera. In Tijuana it is known that those towers catch people. As she was going up the hill, she couldn't breathe because she was so pregnant; she told them she couldn't make it and told them to leave her. The guides told her no, to get up and continue. They stopped and rested at the top of the hill—she couldn't breathe.

The whole time she was army crawling she feared how hard it was on her belly: there were times when her belly was dragging across rocks. They were army crawling and the guide told her, when I say to run, run. They jumped over rocks; they couldn't see anything. She feared rolling down the mountain and losing her baby. She wasn't hungry or thirsty. She had dirt and mud everywhere. There was a large group of people crossing. She saw immigration huts out in the mountains. She was sure they saw her and said, "don't move!" but they didn't charge. While she was crossing and hiding, they would throw themselves to the ground and they would often land on top of each other.

She asked the Virgin, "*que le dije: haznos como hiciste a Juan Diego.*" At this moment, when she prayed, the helicopter turned around, the dogs turned around. "*Este es un milagro.*" When she got to the mountain, she thought she wasn't going to make it, she was praying.

When she relaxed, she realized she had cactus thorns throughout her arms and hands, and she couldn't remove them. When she got to the top and found the thorns on her, they didn't hurt, they didn't hurt as she pulled them up. She looked up to the sky and it was full of stars: "*Cuando miro arriba está lleno de estrellas.*"

They waited for rides underneath rocks. She wasn't cold while they were moving. Once they were under the rock, she found a blanket that had been left from previous crossers.

At this point, they had made it into the United States.

Her baby hadn't moved for hours, and she was worried. There was a little hole in the rock, and she would look up through the hole in the stars and say, "*Por favor, Señor, qué pase un* ride, *que vuelvan por favor.*" And then she heard the cars honk for her, and they ran for the car. They waited at a hotel, but she couldn't eat food. Her baby hadn't moved. She had mud throughout her hair and took a shower, but her baby still hadn't moved. She wasn't hurting from the trip but was just afraid about her baby not moving. She didn't have a ride until the next day, but she couldn't sleep.

One of the other girls was also pregnant and had her small daughter. They would all take turns watching the young child for the pregnant mother. She felt that the men were crybabies, and the women were strong.

A few hours into the hotel stay, her baby started moving and she felt relieved.

When she got to the United States, it was very difficult. They arrived in Los Angeles. She had no family in the United States and living in LA was hard without knowing anyone. In LA, the coyotes hold people until they have family that can come get them or pay them what they're owed. Eventually, she lived in the street when she arrived, and it was difficult. She didn't have parents back in Mexico, but she had siblings that she missed. When they got to Lake Elsinore, life was much easier than Los Angeles.

How would you imagine the borders if you could make it anything you wanted?

"*No hay fronteras perfectas.*" I wouldn't remove the border but make it different. If Mexico asked Americans for visas, if they asked them for papers, maybe they would know how it feels. For Americans to know what it feels like to have the same requirements and rules that we must cross into America. There are a lot of white people from

America and Canada living in Mexico like it's nothing. The *frontera* should be the same but both ways.

I want Americans to know how it feels to be undocumented in another place, as we feel it here. When Americans go to Mexico, they are accommodated in their inability to speak Spanish, they go to tourist areas and speak English only.

How would you imagine or reimagine border patrol?

Just like the US has their wall, and immigration. I would have Americans experience La Migra in Mexico the way Mexicans have to experience it. It's not equal when Americans go over there to Mexico. The border agents lose their humanity when they put people in cages. They need more proper facilities where they detain people, more humane overall.

How would you imagine here, Lake Elsinore, if you could make it anything you wanted?

I would like everyone to be the same despite their skin color. To be treated the same despite our race. They would build more parks for the youth, more programs for kids about to join gangs and to stop the drugs, there are a lot of drugs here. Where there is blight in Lake Elsinore, to put swimming pools and basketball courts. There is a lot of racism among the races. It's a beautiful area to live. Instead of resenting the homeless, we need to help them.

How would you imagine citizenship?

To help immigrants fight for citizenship. To give the parents who don't have papers more assistance. To campaign where all the racist people in the US are taught about their histories to know that they were immigrants too. That the *raza blanca* are not treated better than others.

I would like to work for CPS [Child Protective Services] so I could help people out, because CPS can be rude.

If you could create any technology in the world, what would you create?

A technology for the world, machines that could help remove the contamination of the world, so that our future generations have a better future. Machines with filters that would clean out the air and trash. I would ask for a time machine, but I'm asking for something else more important, which is keeping the earth healthy for our future.

"*Lo más importante ahorita es donde estamos viviendo.*" Two machines: one to break down the trash particles and one machine that cleans the air.

Anything I didn't ask that you would like to say?
I would tell people that want to come to the US to not rush it, to think more about it. Many people come out here and they'll be here for decades and can't get their papers, can't get back to their countries. It's not safe when you cross; they won't know if it will be worth it.

ROSA: I DON'T HAVE LUXURIES, BUT I WORK, AND I FEEL SAFE IN THE UNITED STATES

Rosa crossed through the mountains close to Tijuana. It's a very dangerous crossing, especially across the freeways. If someone fell on the highway, they could get run over. They would see the "mosco" helicopter hovering, and they were told to go back to Mexico. It's dangerous when you cross with a coyote because they could be dangerous too. It's very difficult to come here undocumented, "*Ya sabemos que nada es fácil.*"

She crossed with her two children in November, when La Migra caught them in the mountains. Her son was grabbed, so she gave herself up with her daughter and turned herself in: she didn't want her son to cross alone with a coyote. She had heard of women who lost their kids and didn't want to go through that.

She was in the detention center and would ask immigration for newspaper to lie on, because the cell blocks were pure concrete. In the detention centers they didn't care about the kids at all. This was her experience in 1986.

> Back then there weren't as many cameras. We would run and hide in caves along the mountains. But it was easier back then because the technology today is everywhere. Back then La Migra was on their *motos*, in the helicopter, and on the horses. But now they are watching everything with technology.

They arrived in Lake Elsinore, and she started working. At the time she worked thirteen hours for fifty dollars a day. They saw border patrol around Lake Elsinore often in the years since they've moved there, on the streets or knocking on doors. They seem to target the lower-income apartments.

How would you imagine the borders if you could make it anything you wanted?

I know they must protect the border. There are people crossing to work and others who have other ideas. I know Mexico needs a border but wish there were more chances to cross legally to work in the US.

How would you imagine here, Lake Elsinore, if you could make it anything you wanted?

I wish they would put in more parks, adult schools, to clean the streets better, because we see drugs in the streets. To have more police patrolling certain streets because there are many drug dealers and people doing drugs on the street.

How would you imagine or reimagine border patrol?

I would rather have the border patrol on the border [not in Lake Elsinore]. That's not to be a hypocrite because I'm already here, I do think this country needs to protect itself from people trying to do harm, but some come to work.

If you could create any technology in the world, what would you create?

A technology that eliminates plastic bags. To get rid of plastic bags because they're bad for the earth.

Anything I didn't ask that you would like to say?

In the United States, I don't have problems with anyone. I live well here. I don't have luxuries, but I work, and I feel safe in the US. I like it here.

Conclusion

Esperanza, Yet Hope Remains

In the three years since I've been researching and writing this book, data gathering, and surveillance continues to grow. Virtual and augmented reality can now gather the data of every movement of the body: that data claims to predicts race, gender, and learning abilities through just a few motions (Lambert, 2018). Clearview AI continues to exploit data privacy, with recent news that they will be selling their products to Latin America and Asia (Winter, 2021).

But something else has happened too: activism for data and human rights has increased and shows up where we often don't expect. Tech workers are walking off their jobs over unethical practices, new technology collaborations by and for BIPOC are developing, and more theorizing has arisen to ask if information and technologies are tools or hindering liberation. To conclude this book, I want to discuss responses to this state of data borders, and offer that hope, *esperanza,* is always already present.

In this conclusion, I look at advocates for change, for immigrant information and data rights to disrupt the bigger story structures that define and improve tech surveillance, detention, and deportation.

CHANGE THE SUBJECT: CHALLENGING THE POWER TO NAME

Library and information science scholar Hope Olson (1998) demonstrates that how information is organized reflects the larger social naming schemas: "classificatory structures are developed by the most powerful discourses in a society. The result is the marginalization of concepts outside the mainstream" (p. 235). The ways in which information is gathered, classified, organized, stored, correlated, and retrieved exerts normative structures of power through *naming*. Noble reminds us that "search engines, like other databases of information, are equally bounded, limited to providing only information based on what is indexed within the network" (2018, p. 142). Unfortunately, because private companies have proprietary rights over our data, the ways in which they classify it are often cloaked. Noble's analysis of Google applies just as well to Palantir, Amazon, Anduril, and more Silicon Valley–related private companies : "because it is a commercial enterprise, the discussions about its similar information practices are situated under the auspices of free speech and protected corporate speech, rather than being posited as an information resource that is working in the public domain, much like a library" (p. 143). So public policy leading with immigrant data rights would demand more transparency of those tech companies that hold government contracts with ICE.

We can glean organizing methods for data schemas from what has happened within and around the library, as an example of challenging those naming schemas and challenging *who* gets to name and classify themselves in data organizing.

In 2016, Latinx college students at Dartmouth University challenged the Library of Congress's subject heading Illegal Aliens used in cataloging and metadata to describe undocumented people, immigrants, and Latinx residents in general. The term *illegal aliens* was used to categorize books, articles, academic records, and so on. As information is classified with that term, data is organized, stored, and retrieved according to that language: "To see a word that starkly represents one way to dehumanize undocumented people emotionally hurt me, but also did make me not really trust the library as a system," said Dartmouth University student Melissa Padilla (Broadley & Baron, 2019).

Students organized meetings with libraries and university administration, protests and marches, and a campaign to petition for the change at the Library of Congress (LC). Students Melissa Padilla and Óscar Rubén Cornejo Cásares, along with Dartmouth librarian Jill Baron, petitioned the LC to change the subject heading to Undocumented Immigrants. That petition was rejected in 2014. In 2014 many librarians advocated for support of replacing the LC subject heading, and the American Library Association endorsed the students' petition. In March

2016, the LC announced that they would delete the subject heading Illegal Aliens In November 2021, the LC finally replaced the Illegal Immigration and Aliens subject heading, with headings on existing bibliographic records in the catalog being updated immediately.

What followed demonstrated the importance of the power to name in knowledge classification schemas: Representative Diane Black (R–TN) blocked the LC subject heading change through legislation. Republican media and congresspeople claimed that it was just a language change for "political correctness," saying that the former term was historic. Following that backlash, the LC changed the subject heading back. However, individual institutions, such as the University of California library system, have since changed that subject heading individually (Broadley & Baron, 2019). Those Republican lawmakers were especially troubled by a movement to change language, signaling that language should be unchanged and fixed.

What we learn from Dartmouth students, librarians, and professors is that there must be an activist arm to implement the power to name in data classification. And for the purposes of the technology used by ICE, that would mean more transparency in how data is organized and classified around surveilling, detention, and deportation of immigrants. Dartmouth students also call for an adaption to *change*, a change to language as it classifies people in information systems.

Data transparency and the ability to see how Latinx people are classified and organized within information systems is crucial. Latinx college students saw an injustice within data classification and named it out of their own contextualized experience. They engaged with what Catherine D'Ignazio and Lauren F. Klein have named *data feminism:* "a way of thinking about data, both their uses and their limits, that is informed by direct experience, by a commitment to action, and by intersectional feminist thought" (2020, p. 8).

Change within the field of library and information science (LIS) can lend itself to the bigger picture of technology design around Latinx immigrants. Classification schemas are one of the primary methods of organizing contemporary technologies (Crawford, 2021). Olson has long warned of the dangers of simplistic classification schemas. However, Olson also warns about the responses or fixes to classification systems. Rather than inserting fixes into existing AI to be all inclusive, a response would be for everyone involved in designing classification systems to take accountability for their actions. When asked about her book *Algorithms of Oppression* and the response from the technology industry, Noble notes that tech companies responded to her book as a series of tickets in their customer service system, or tech errors that they could fix, rather than recognizing that structural racism and sexism is built into information systems (Noble, 2021). These systems must be viewed as deeply embedded in structural hierarchies, rather than fixes. The lesson, then, from this case is not just that language can and should change in machine learning. But also, the lesson is their *method* in organizing.

THE CHANGING IMMIGRANT DATA RIGHTS MOVEMENT

The right to data privacy and the immigrant rights movement have also changed and coalesced over the years. Increasing numbers of tech workers are speaking up, walking out, and protesting their own workplace's unethical data practices.

Perhaps one of the most resounding recent activist events in the fields of AI and technology was the 2018 series of Google Walkouts by Google employees and the firing of Timnit Gebru. Gebru, the cofounder of Black in AI, was fired from Google after she questioned their commitment to diversity and coauthored a paper with fellow Google employees about how AI can lead to bias through large data sets. Gebru said: "Academics should not hedge their bets but take a stand. . . . This is not about intentions. It's about power, and multinational corporations have too much power and they need to be regulated" (Johnson, 2021).

Many people find themselves enmeshed in the ICE information deportations that go beyond technology industries. In 2019 hundreds of employees working at the Boston online furniture company Wayfair walked off their jobs when the company signed a contract to provide furniture for the inhumane immigrant detention centers located around the United States (DeCosta-Klipa, 2019). Citizens and permanent residents find themselves deep into the lucrative business of undocumented immigrants' detention and deportation. A web of information labor all over the world delivers digital tools into our hands, and those tools are networked into ICE databases.

Black in AI and Queer in AI outline the defunding approach, which is used increasingly among tech workers: "What does it mean to have ethical funding? What kinds of organizations should fund a conference?" (Johnson, 2021). The California Academic & Research Libraries Association (CARL, 2020) has a written policy that it will not allow vendor support at conferences if those vendors contribute to unethical practices that further lead to the marginalization of communities.

There are activist groups for data privacy and human rights that have emerged from common affinities and overlap with immigrant rights movements. The National Immigration Project, the Immigrant Defense Project, and Mijente drafted a comprehensive report, *Who's behind ICE*, on the tech and data companies invested in ICE (National Immigration Project et al., 2019). Mijente has a NoTechforICE.com campaign that mobilizes people across the country to advocate against immigrant data privacy violations. These groups advocate for policy change, organize student protests and boycotts around the country, and played a major role in enlisting Latinx voters in the 2020 presidential elections. Mijente advocates that sanctuary cities now must include the protection of immigrant data rights and privacy through policy (Mijente, 2017).

The Electronic Frontier Foundation (EFF, n.d.) is a nonprofit that defends digital privacy and include a Surveillance Self-Defense online guide for "defending yourself and your friends from surveillance by using secure technology and developing careful practices." EFF has responded to several immigrant data rights violations including when the Department of Homeland Security announced they would use social media surveillance in "extreme vetting" of immigrants.

The Our Data Bodies Project has published a Digital Defense Playbook in English and Spanish as a guide to knowing your data rights (Lewis et al., 2018). The Our Data Bodies Project advocates that data bodies can be a source of marginalization and agency, stating that *data* is part of people's stories: "[W]e must first understand how we are both hurt and helped by data-based technologies. This work is important because our data is our stories. When our data is manipulated, distorted, stolen, or misused, our communities are stifled, and our ability to prosper decreases" (Our Data Bodies Project, n.d.). Our Data Bodies offers guides in building agency through know-your-data-rights tools; and advocating against surveillance in its detrimental use against BIPOC, LGBTQ+, and lower income communities; but also teaches that systems and digital initiatives can build positive outcomes and innovation.

Data for Black Lives is an organizing group that organizes around data rights with the goals of abolishing big data, advocacy, organizing initiatives and cohorts, and policy working groups. This group also advocates for Facebook to give Black people access to their own data and for COVID-19 race data to be transparent state by state (Data for Black Lives, n.d.).

We also see affinity groups in professions that are impacted by immigrant surveillance organizing around data privacy rights. For example, The Library Freedom Project advocates for communities' privacy from a social justice, feminist, and anti-racist perspective. The project has crash courses and programs to train librarians and LIS professionals on systems and policies around vendor agreements, conducting privacy audits, and working with library IT to ensure data privacy (Library Freedom Project, n.d.). One method is policy that divests from companies that collaborate with ICE. As mentioned previously, CARL built a code of ethics into their conference sponsorship requirements that states sponsors must "respect the privacy and humanity of our users, and do not engage in the practice of selling information to law enforcement" (California Academic & Research Libraries, 2020).

We can also learn from the models for equity of contemporary social movements. Adrienne maree brown's method of emergent strategy considers organizing principles for social change and has tools for organizing with social justice methods. Dean Spade's work looks at mutual aid in times of crisis, where organizing for equity isn't done for purposes of profit or charity but around affinities that work towards liberatory projects (Spade, 2020). These methods are already employed by these groups and can be more intentionally applied to the fields of LIS and STEM fields to counter technologies that increase oppression.

CHANGE POLICY

Over the past few years, the level of surveillance, big data gathering, and financial growth in data brokering have come to the forefront of American consciousness. When I began writing this work, I wanted to make visible some of the systems that were building data borders in every part of the United States and every aspect of our lives. But in the past years, as COVID-19 has overturned our lives and action for racial equity has become more widespread, the discussion around the data body has become more visible and prevalent as well. For many tech companies, it's a race to the bottom in data privacy and data gathering, while upholding the semblance of democratic values. The state of surveillance has become dystopic, and consumers are more conscious of this dystopia. Facebook is leading with their own policy campaign for internet regulation in which they set the terms (The Daily, 2021). Apple is demonstrating that US consumers can opt out of data brokering, rather than submitting to any of that pesky internet regulation (The

Daily, 2021). But they continue to share consumer data in China and to allow Chinese government censoring of the media.

Policy in favor of data rights must precede the egregious violations that we learn about every day, and it must lead with marginalized communities' data privacy rights. For example, to advocate that the Senate bills that prioritize data privacy protect citizens *and* undocumented people from borderland surveillance, we must prioritize undocumented people's biodata rights in the United States. As discussion around antitrust and evolving internet policies arises, Sarah Lamdan (2022) asks us to hold RELX and Thomson Reuters to the same standards in working against monopolies as we do with other major Silicon Valley companies such as Google, Facebook, and Palantir.

We will continue to see policy that seeks data rights for citizens at the exclusion of undocumented immigrants, or that uses undocumented immigrants as the exception to the rule. We must organize data privacy policy rights with attorney and transgender rights organizer Dean Spade's trickle-up social justice approach, from his 2009 lecture at Barnard College: "We have to ask ourselves does this divide our community by leaving out vulnerable people. . . . [A] lot of the rights people have been fighting for only benefit the very few. They usually harm the people that are at the bottom . . . based on a commitment that social justice doesn't trickle down. We should center the experiences of the most vulnerable first, that's how we should determine our agenda . . . " (Spade, 2009).

Not taking a trickle-up social justice approach and only discussing data privacy policy from citizens' rights will not benefit undocumented people, whereas leading with undocumented people's, DACA recipients', and permanent residents' data privacy rights will also benefit citizens.

Another approach to changing the ways that we think about information technologies is to listen to the critical information, science, and technology scholars' and industry whistleblowers' forewarnings about the surveillance dragnet to come. We should include critical information and technology theories into STEM-based programming as a mandatory part of curriculum. Adopting critical theories into required STEM curriculum, such as programs that teach big data methods, will assist in including ethical compasses alongside technology skills.

CHANGE THE ETHICS AROUND DATA COLLECTION

In her 2021 keynote address at the Northwest Archives conference, archivist Tracy Drake reflected on what it meant to preserve and digitize archives in crisis moments of 2020–21. Her keynote "We Are Each Other's Business: Archiving with Intention in Ever Changing Times" considered the ethics of care regarding such important but traumatizing information work, to reimagine the future of archives in LIS. Her title was a quote from the late great Black poet Gwendolyn Brooks's 1970 poem *Paul Robeson:* "We are each other's harvest / we are each other's business / we are each other's magnitude and bond" (Brooks, 2005). Drake is thinking about information work in a real-time crisis. The process of archiving often includes digitizing work, generating metadata for that work, classifying sensitive and often traumatizing racial histories and present-day memories as they're happening, and making that information retrievable for patrons. It's work not without its own trauma:

> Our duty of care extends to the communities we serve, i.e., our donors' research communities, and the communities where our institutions reside. As archivist for the college, my duty of care extends to the campus community, including faculty, staff and students. . . . As archivists placing our selves as members within the broader community, recalibrate our approach to the acquisition of collections, moving us away from the colonial mindset of extracting resources, artifacts, and collections from communities to one with a shared value vision and respect, leading to less harmful collecting practices. This means we must be active members of the communities and respond to these needs in real time. (T. Drake, personal communication, May 10, 2021)

Drake asks her field for an ethics of care when documenting people, especially marginalized people. Her approach inverts that of data borders and data bodies. Rather than archiving the mass data gathering by commercial and military technology companies, she considers the community and asks about the *context* in which that information takes place. In reflecting on racism and sexism embedded in algorithms, Noble advocates for giving information context as well: "It is this point that I want to emphasize in the context of information retrieval: information provided to a user is deeply contextualized and stands within a frame of reference. For this reason, it is important to study the social context of those who are organizing information and the potential impacts of the judgements inherent in informational organization processes. Information must be treated in a context" (2018, p. 149).

Drake outlines a manifesto for an ethics of care and radical empathy that applies technology design, data management, data storage, and retrieval in big data and is viewed as furthering this project of giving information context. The manifesto she spoke of in her keynote serves as an alternative to contemporary surveillance models built by Silicon Valley:

> We have a responsibility to collect, document, and archive narratives that go beyond trauma, but also explore joy because people and communities are not one dimensional. . . .
>
> We have a responsibility to the communities with which we reside to act with intention and integrity. I want to take a second and reiterate an important word I just used in the previous sentence, responsibility, because I think it's critical to help us understand the answer to that question: What do we owe to marginalized communities?
>
> We have a responsibility to not act on biases because of those that appear in the historical record. In this instance, we must provide information around problematic and offensive language in our finding aids and use inclusive and contemporary language that makes information easily retrievable.
>
> And finally, we must document narratives from various perspectives and not just the organizations or individuals in power. (T. Drake, personal communication, May 10, 2021)

We have to imagine alternative futures to respond to this state of data borders and the increasing surveillance state where bodies are organized by data, instead giving information context. We must lead with an ethics of care and radical empathy in data activism and policy organizing, technology design, data gathering, metadata classifications, correlation, big data, and AI. Rather than building communities out of correlating data for purposes that surveil, Drake drives home the person- and community-first approach to data gathering: that we are each other's business.

GIVING THE DATA BORDER CONTEXT

I began this book with my own experience straddling multiple borderlands: geographic, interpersonal, and cultural. This book argues that most US residents are already engaged in a data borderland state. People's consciousness of participation in this state varies. It's a state in which information and data bodies live together, out there, beyond us. And one that *feels* beyond description. Present and absent. Our data is proprietary but not our own property.

The cloaked nature of surveillance—both present and formless, such as Amazon Alexa's and Amazon Web Services' data storage cloud—is a signature of these data borders. The uncertainty about when and who is networked into detention and deportation is also a signature of data borders.

I hope that this book works as a parable, in that it can change mainstream ways of thinking about data rights and technology to consider that (1) immigrants deserve data rights equal to those of the most privileged of citizens, and (2) that technology does not lead to a direct trajectory of positive progress.

Growing up, when the border patrol cars and vans were out, I observed a network of women text each other about their last seen location. Men left their work meeting spots and skipped that day's wages. Their landscaping tools standing at attention in pickup trucks were not to be used that day. Moms chose not to drive their kids to school.

It was common to see the border patrol's white-and-green van as we drove south on I-15. My mom would say a quick prayer for wherever they were going. A *bendición* for the shadowed outlines inside the tinted windows.

"There are people in there," my mom would say.

There are people in there. My arms tingled.

If there's one thing that I hope you take away from this book, it's a feeling. When you buy something on Amazon, or see Palantir and Anduril logos, or when you search RELX or Thomson Reuters databases to research and write a paper, may the hairs on your arms rise and you sense the data borderland in which you reside. May you feel that there are people in there.

Acknowledgments

No one writes a book alone, and with every project I'm so grateful for my community's support. My mom and dad have been lifelong supporters and cheerleaders of my work. Thank you to my partner Julio Cesar and Xochitl for support, laughter, and giving me a full life. Matt and Sofie's presence and fun during our time together sustains me. Thank you to the Villa, Nicholas, and Arias-Oropeza families for your support even when we live so far apart.

My work could never be possible without the various forms of childcare and day care necessary for parents to excel in academia. I'm grateful for all the caregivers and educators who keep my daughter safe, fed, rested, and playing so that I can work with peace of mind.

This research wouldn't be possible without funding from the University of Rhode Island. The following sources funded this research: the University of Rhode Island's Project Completion Grant, the A&S Dean's Grant, the A&S Proposal Development Grant, and funds from the Graduate School of Library and Information Studies (GSLIS) at URI. Thank you to URI's Social Science Institute for Research, Education, and Policy (SSIREP) grant for funding the next part of this project. Parts of this book were previously published and have been reprinted with permission from *Bitch Media* and *Feminist Media Studies* on Taylor & Francis Online.

Receiving feedback from strong reviewers is an invaluable component to writing. I am so grateful to my three reviewers who took the time and thoughtfulness to improve this book. Thank you to the staff at UC Press, especially my wonderful editor Michelle, editorial assistant LeKeisha, and associate editor Enrique.

My group of friends act as a sounding board, reflect, challenge, laugh, cry, fail, grieve, and celebrate successes. The mentors, colleagues, and friends to whom I am deeply indebted include Dr. Jeanie Austin, Aisha Connor Gaten,

Tracy Drake, Dr. Mar Hicks, Dr. Safiya Noble, Dr. Colin Rhinesmith, Dr. K. R. Roberto, Dr. Sarah T. Roberts, Dr. Miriam Sweeney, and Dr. LaTesha Velez. A special thanks to Sarah Lamdan for her consultation on this project.

My lifelong friends have enriched my life: Danielle, Mikah, Jennie, Bethy, Kim, Nicole, Laura, Sara, and Crystal for providing such wonderful friendships, even when we are states apart.

I am fortunate for colleagues at URI who been very supportive over the years. Thank you to my colleauges at GSLIS, Dr. Kendall Moore, Dr. Scott Kushner, Dr. John Pantone, and Dr. Julie Keller, as well as my dean in the administration who continues to support and fund my research and travel and believe in my vision for scholarship and education. I'm so grateful for support staff at GSLIS Jessica Nalbandian and Cassandra O'Brien. My students have helped me grow over the years in thinking about critical information science. I'm especially grateful to Kathleen Fieffe, Miranda Dube, Kate Fox, Jo Knapp, Christina Swiszcz, and Olivia Phillips, along with all of my students that make the field of LIS excellent.

Many mentors and supportive faculty got me to where I am today from undergraduate through my doctoral program, and I'm always indebted to their support during the many years of my education. I am indebted to Professor Lynda Barry, Dr. Young Lee Hertig, Dr. Christine Pawley, Dr. Linda Smith, Dr. Angharad Valdivia, Dr. Sharra Vostral, and Dr. Ethelene Whitmire.

And finally, thank you so much to my interview collaborators in my hometown and around Southern California.

References

Acxiom. (2018). 2018 annual report. https://www.annualreports.com/HostedData/AnnualReports/PDF/NASDAQ_ACXM_2018.pdf

Alden, W. (2016, May 6). Inside Palantir, Silicon Valley's most secretive company. *Buzzfeed News*. https://www.buzzfeednews.com/article/williamalden/inside-palantir-silicon-valleys-most-secretive-company

Allen, M. (2020, May 25). *AXIOS on HBO: Palantir CEO Alex Karp on work for ICE* [Film]. YouTube. https://www.youtube.com/watch?v=ChwSTuDa9RY

American Civil Liberties Union. (n.d.). *The constitution in the 100-mile border zone*. https://www.aclu.org/other/constitution-100-mile-border-zone

American Civil Liberties Union. (2022). Mass incarceration. https://www.aclu.org/issues/smart-justice/mass-incarceration

American Defamation League. (2014). *Longtime anti-immigrant activists behind Murrieta protests*. https://www.adl.org/news/article/longtime-anti-immigrant-activists-behind-murrieta-protests

ANDE. (n.d.). About. ANDE. Retrieved December 1, 2019, from https://www.ande.com/about-ande-rapid-dna/

ANDE. (n.d.). What is rapid DNA? ANDE. Retrieved December 1, 2019, from https://www.ande.com/about-ande-rapid-dna/

Anderson, B. (1991). *Imagined communities: Reflections on the origin and spread of nationalism*. Verso Press.

Anderson, C. (2008, August). The end of theory: The data deluge makes the scientific method obsolete. *Wired*. https://www.wired.com/2008/06/pb-theory/

Anderson, E. (2014, May 24). The evolution of electronic monitoring devices. NPR. https://www.npr.org/2014/05/22/314874232/the-history-of-electronic-monitoring-devices

Andrejevic, M. (2007). Surveillance in the digital enclosure. *The Communication Review 10:4*, 295–317.

Anduril Industries. (n.d.). Lattice OS. Retrieved January 15, 2020, from https://www.anduril.com/lattice

Anduril Industries. (n.d.). Mission aligned. Retrieved August 15, 2020, and November 29, 2022, from https://www.anduril.com/company

Anduril Industries. (n.d.). Our work. Retrieved January 15, 2020, from https://www.anduril.com/work

Anzaldúa, G. (1987). *Borderlands: La frontera.* Aunt Lute Press.

Austin, J. (2020). Information access within carceral institutions. *Feminist Media Studies, 20(*8),1293–1297.

Barrington, I. (2015, November 24). How railroad history shaped internet history. *The Atlantic.* https://www.theatlantic.com/technology/archive/2015/11/how-railroad-history-shaped-internet-history/417414/

Baynes, C. (2019, June 14). Government "deported 7,000 foreign students after falsely accusing them of cheating in English language tests." *Independent.* https://www.independent.co.uk/news/uk/politics/home-office-mistakenly-deported-thousands-foreign-students-cheating-language-tests-theresa-may-windrush-a8331906.html

Bawden, D., & Robinson, L. (2013). Introduction to information science. Neal-Schuman.

BBC. (2019, December 16). Top tech firms sued over DR Congo cobalt mining deaths. https://www.bbc.com/news/world-africa-50812616

Bedoya, A. (2020, September 22). The cruel new era of data-driven deportation. *Slate.* https://slate.com/technology/2020/09/palantir-ice-deportation-immigrant-surveillance-big-data.html

Bell, D. (1979). The social framework of the information society. In M. Dertouzos and J. Moses (Eds), *The computer age: A twenty-year view.* (pp. 163–211). MIT Press.

Benjamin, R. (2019). *Race after technology.* Polity Press.

The Blackivists. (2020, June 2). The blackivists' five tips for organizers, protestors, and anyone documenting movements. *Sixty inches from center.* https://sixtyinchesfromcenter.org/the-blackivists-five-tips-for-organizers-protestors-and-anyone-documenting-movements/

Boyd, D., & Crawford, K. (2012). Critical questions for big data: Provocations for a cultural, technological, and scholarly phenomenon. *Information, Communication & Society.* 15(5), 662–679.

Boyle, J. (2003). The second enclosure movement and the construction of the public domain. *Law and Contemporary Problems,* 66(1/2), 33–74.

Bouk, D. (2016). How our days became numbered: Risk and the rise of the statistical individual. University of Chicago Press.

Boyle, G. (2021). *The whole language: The power of extravagant tenderness.* Simon & Schuster.

Buolamwini, J., & Gebru, T. (2018). Gender shades: Intersectional accuracy disparities in commercial gender classification. In Proceedings of the 1st Conference on Fairness, Accountability and Transparency, *PMLR,* no. 81, 77–91. http://proceedings.mlr.press/v81/buolamwini18a/buolamwini18a.pdf

Broadley, S., & Baron, J. (2019). *Change the subject* [Film]. Dartmouth Digital Library Program.

Brooks, G. (2005). *The essential Gwendolyn Brooks*. Library of America.
Brown, a. m. (2018). *Emergent strategy: Shaping change, changing worlds*. AK Press.
Brown, B. (2015). Rising strong: How the ability to reset transforms the way we live, love, parent, and lead. Spiegel & Grau.
Brown, B. (2021, September 22). Partnerships, patters, and paradoxical relationships with Esther Perel [Audio podcast episode]. In *Unlocking us*. https://brenebrown.com/podcast/partnerships-patterns-and-paradoxical-relationships/
Brown, S. (2015). *Dark matters: On the surveillance of blackness*. Duke University Press.
Bump, P. (2016, August 30). The last time the United States tried to build a virtual border wall, it wasn't exactly a big success. *The Washington Post*. https://www.washingtonpost.com/news/the-fix/wp/2016/08/30/the-last-time-the-united-states-tried-to-build-a-virtual-border-wall-it-wasnt-exactly-a-big-success/
Bundy, A. (2019, April 26). A virtual wall may be the solution to protect wildlife at the border. *Cronkite News*, Arizona PBS. https://cronkitenews.azpbs.org/2019/04/26/us-mexico-border-lidar/
Burrington, I. (2015, November 24). How railroad history shaped internet history. *The Atlantic*. https://www.theatlantic.com/technology/archive/2015/11/how-railroad-history-shaped-internet-history/417414/
Butler, O. (2019). *Parable of the Sower: A powerful tale of a dark and dystopian future* (rev. ed.). Headline Press.
Butler, O. (2012). *Parable of the sower*. Open Road Media Sci-Fi & Fantasy.
Cacho, L. M. (2012). *Social death: Racialized rightlessness and the criminalization of the unprotected*. New York University Press.
Cagle, M. (2017). Why are border sheriffs rushing to adopt recognition technology? ACLU. https://www.aclu.org/blog/privacy-technology/surveillance-technologies/why-are-border-sheriffs-rushing-adopt-iris
California Academic & Research Libraries Association. (2020). Sponsorship requirements. https://conf2020.carl-acrl.org/sponsorships/
California State Auditor. (2016, August). The CalGang criminal intelligence system. https://www.auditor.ca.gov/pdfs/reports/2015–130.pdf
Campbell, H. (2017, October). Dr. Robert Burton on being certain. *Human Current*. http://www.human-current.com/blog/2017/10/4/dr-burton-on-being-certain
Campbell, J. (1968). The hero with a thousand faces. Princeton University Press (Original work published 1949)
Campbell, J. (1988). *Joseph Campbell and the power of myth. Interviews by Bill Moyers*. Doubleday & Company.
Carlson, N. (2010, September 21). The Facebook movie is an act of cold-blooded revenge—new, unpublished IMs tell the real story. *Business Insider*. https://www.businessinsider.com/facebook-movie-zuckerberg-ims
Castells, M. (1997). An introduction to the information age. *City*, 2(7), 6–16.
Chaar-López, I. (2019). Sensing intruders: Race and the automation of border control. *American Quarterly*, 71(2), 495–518. doi:10.1353/aq.2019.0040.

Chabram-Dernersesian, A. (2006). Introduction to Part One in *The chicana/o cultural studies reader* edited by Angie Chabram-Dernersesian. Routledge.

Chase, P., Bailey, B, Nocek, J., & Christian, G. (2016, September 28). Connecticut four reunite against FBI overreach. *American Libraries*. https://americanlibrariesmagazine.org/blogs/the-scoop/connecticut-four-librarians-fbi-overreach/

Chavez, L. (2008). *The Latino threat: Constructing immigrants, citizens, and the nation*. Stanford University Press.

Chen, M. (2019, October 6). The US border security industry could be worth $740 billion by 2023. *Truthout*. https://truthout.org/articles/the-us-border-security-industry-could-be-worth-740-billion-by-2023/

Christl, W. (2017). Corporate surveillance in everyday life. Cracked Labs. https://crackedlabs.org/en/corporate-surveillance

Cifor, M., Garcia, P., Cowan, T. L., Rault, J., Sutherland, T., Chan, A., Rode, J., Hoffmann, A. L., Salehi, N., & Nakamura, L. (2019). Feminist data manifest-No. Retrieved December 2, 2022, from https://www.manifestno.com/

Cioffi-Revilla, C., & Rouleau, M. (2010). MASON RebeLand: An agent-based model of politics, environment, and insurgency. *International Studies Review, 12(1)*, 31–52.

CNN. (2016, October 19). Donald Trump: We need to get out "bad hombres." [Film]. YouTube. https://www.youtube.com/watch?v=AneeacsvNwU

Condis, M. (2018). *Gaming masculinity: Trolls, fake geeks, and the gendered battle for the online culture*. University of Iowa Press.

Congressional Research Service. (2020, September 24). Facial recognition technology and law enforcement: Select constitutional considerations. US Congress. https://crsreports.congress.gov/product/pdf/R/R46541

Coolfire Solutions. (2018). A digital wall could be the answer to U.S. border security. https://www.coolfiresolutions.com/blog/mexico-us-border-security-digital-wall/

Coolfire Solutions. (n.d.). What is situational awareness? Retrieved July 20, 2022, from https://www.coolfiresolutions.com/blog/what-is-situational-awareness/

Corrigan, J. (2019, August 20). DHS is collecting biometrics on thousands of refugees who will never enter the U.S. *Nextgov*. https://www.nextgov.com/emerging-tech/2019/08/dhs-collecting-biometrics-thousands-refugees-who-will-never-enter-us/159310/

Cowie, J. (2001). *Capital moves: RCA's seventy-year quest for cheap labor*. New Press.

Crawford, K. (2021). *Atlas of AI: Power, politics, and the planetary costs of artificial intelligence*. Yale University Press.

Cron, L. (2012). *Wired for story: The writer's guide to using brain science to hook readers from the very first sentence*. Ten Speed Press.

Cron, L. (2014, May 14). *Wired for Story*. TEDx Talks. https://www.youtube.com/watch?v=74uvomJSouM

Cuevas, S. (2010, July 28). Temecula-area cities adopt anti-illegal immigration laws. *Southern California Public Radio*. https://www.scpr.org/news/2010/07/28/17802/temecula-area-cities-adopt-anti-illegal-immigratio/

Currier, C. (2019, November 14). Lawyers and scholars to LexisNexis, Thomson Reuters: Stop helping ICE deport people. *The Intercept.* https://theintercept.com/2019/11/14/ice-lexisnexis-thomson-reuters-database/

The Daily. (2021, May 11). Apple vs. Facebook. [Audio podcast episode]. *New York Times.* https://www.nytimes.com/2021/05/11/podcasts/the-daily/mark-zuckerberg-tim-cook-facebook-apple.html

Data for Black Lives. (n.d.). About D4BL. https://d4bl.org/about

Dave, P. (2022, March 30). Google cuts racy results by 30% for searches like "Latina teenager." Reuters. https://www.reuters.com/technology/google-cuts-racy-results-by-30-searches-like-latina-teenager-2022–03–30/

Davis, K. (2019, March 23). How would a "smart wall" work at the U.S.-Mexico border? *Los Angeles Times.* https://www.latimes.com/local/lanow/la-me-ln-smart-border-wall-20190324-story.html

Dean, S. (2019, July 26). A 26-year-old billionaire is building virtual border walls—and the federal government is buying. *Los Angeles Times.* https://www.latimes.com/business/story/2019–07–25/anduril-profile-palmer-luckey-border-controversy

DeCosta-Klipa, N. (2019, June 26). Here's what happened during the Wayfair walkout in Boston. *Boston.* https://www.boston.com/news/local-news/2019/06/26/wayfair-walkout-boston

Department of Homeland Security. (2015, November 24). Silicon Valley office fact sheet. https://www.dhs.gov/publication/silicon-valley-office

Department of Homeland Security. (2019, November 5). Snapshot: U.S. border patrol agents leverage emerging S&T tech to ensure the security of our nation's borders. https://www.dhs.gov/science-and-technology/news/2019/11/05/snapshot-us-border-patrol-agents-leverage-emerging-st-tech

D'Ignazio, C., & Klein, L. F. (2020). *Data feminism.* MIT Press.

Dinerstein, J. (2006). Technology and its discontents: On the verge of the post-human. *American Quarterly,* 58(3), 569–595.

Donovan, B. (2015, January 15). How much is consumer data worth? It depends! Acxiom.com. Retrieved December 1, 2022, from https://www.acxiom.com/blog/how-much-is-consumer-data-worth/

Durbin, D. (2019, January 18). U.S. Sen. Dick Durbin: Let's end this shutdown and work on a smart border security plan, not a wall. *Chicago Tribune.* https://www.chicagotribune.com/opinion/commentary/ct-perspec-dick-durbin-trump-border-security-wall-shutdown-0120–20190117-story.html

Edwards, P. N. (1997). *The closed world: Computers and the politics of discourse in Cold War America.* MIT Press.

Eglash, R. (2002). Race, sex, and nerds: From Black geeks to Asian American hipsters. *Social Text. 71*(20), 49–64.

Electronic Frontier Foundation. (n.d.). Surveillance self-defense: Tips, tools and how-tos for safer online communications. https://ssd.eff.org/en

Elias, M. (2013). The School-to-Prison Pipeline: Policies and practices that favor incarceration over education do us all a grave injustice. *Teaching Tolerance.* Southern Poverty Law Center. https://www.tolerance.org/magazine/spring-2013/the-school-to-prison-pipeline

Encyclopaedia Britannica. (n.d.). Bill Gates | American computer programmer, businessman, and philanthropist. Retrieved December 7, 2018, from https://www.britannica.com/biography/Bill-Gates

Erazo, E. (1995) . Information highway update. *REFORMA Newsletter, 13*(4), 16.

Esposito, R. (2010). *Communitas: The origin and destiny of community*. Stanford University Press.

Esquivel, P. (2019, June 28). Detainees held for prolonged periods at temporary California facility, Border Patrol says. *San Francisco Examiner.* https://www.sfexaminer.com/national-news/detainees-held-for-prolonged-periods-at-temporary-california-facility-border-patrol-says/

Etter, L., & Weise, K. (2018, April 9). America's virtual border wall is a 1,954-mile-long money pit. *Bloomberg*. https://www.bloomberg.com/news/features/2018–04–09/the-race-to-cash-in-on-trump-s-invisible-high-tech-border-wall?leadSource=uverify%20wall

Eubanks, V. (2017). *Automating inequality: How high-tech tools profile, police, and punish the poor*. St. Martin's Press.

Fang, L. (2019, March 9). Defense tech startup founded by Trump's most prominent silicon valley supporters wins secretive military ai contract. *The Intercept*. https://theintercept.com/2019/03/09/anduril-industries-project-maven-palmer-luckey/

Farberov, S., & Associated Press. (2014, July 15). Protestors both pro and anti-immigration descend on small town as they battle over transfer of undocumented children. *Daily Mail*. https://www.dailymail.co.uk/news/article-2692559/Arizona-protesters-hope-stop-immigrant-transfer.html

Fojas, C. (2021). *Border Optics: Surveillance Cultures on the US-Mexico Frontier*. New York University Press.

Forbes. (2016, December 12). #22 Palmer Luckey. Retrieved August 31, 2020, from https://www.forbes.com/profile/palmer-luckey/#39ff4adc1409

Funari, V., & De La Torre, S. (2006). *Maquilapolis* [Film]. California Newsreel.

Funk, M. (2019, October 2). How ICE picks their targets. *New York Times*. https://www.nytimes.com/2019/10/02/magazine/ice-surveillance-deportation.html

Gable, R.S. (2015). The ankle bracelet is history: An informal review of the birth and death of a monitoring technology. *The Journal of Offender Monitoring, 27*(1), 4–8. https://www.civicresearchinstitute.com/online/PDF/The%20Ankle%20Bracelet%20Is%20History.pdf

Gamboa, S. (2020, November). Rise in reports of hate crimes against Latinos pushes overall numbers to 11-year high. NBC News. https://www.nbcnews.com/news/latino/rise-hate-crimes-against-latinos-pushes-overall-number-highest-over-n1247932

Garcilazo, J.M. (2012). *Traqueros: Mexican railroad workers in the United States, 1870–1930* (1st ed.). University of North Texas Press.

GCN Staff. (2019, November 25). Digital sand table helps CPB visualize terraincreates a 3D map of any given terrain so agents can gain situational awareness ahead of missions. *GCN*. https://gcn.com/emerging-tech/2019/11/digital-sandtable-helps-cbp-visualize-terrain/297694/

George, J. (2017, July). The biometric frontier: "Show me your papers" becomes "Open your eyes" as border sheriffs expand iris surveillance. *The Intercept*. https://theintercept.com/2017/07/08/border-sheriffs-iris-surveillance-biometrics/

Ghaffary, S. (2020, February 7). The "smarter" wall: How drones, sensors, and AI are patrolling the border. *Vox*. https://www.vox.com/recode/2019/5/16/18511583/smart-border-wall-drones-sensors-ai

Glassdoor contributor. Save the shire. (2014, November 5). Glassdoor. Retrieved December 1, 2021, from https://www.glassdoor.co.uk/Reviews/Employee-Review-Palantir-Technologies-RVW5318457.htm

Goldberg, D., Wilson, J., & Knoblock, C. (2007). From text to geographic coordinates: The current state of geocoding. *Journal of Urban and Regional Information Systems Association, 19*(1), 33–46.

Gómez-Peña, G. (2001). The virtual barrio @ the other frontier: (Or the chicano interneta). New York University Press.

Gottschall, J. (2013). *The storytelling animal: How stories make us human*. Mariner Books.

Gottschall, J. (2014, May). *The storytelling animal* [Film]. TED Conferences. https://www.youtube.com/watch?v=VhdoXdedLpY

Gray, M. & Suri, S. (2019). *Ghost work: How to stop Silicon Valley from building a new global underclass*. Houghton Mifflin Harcourt.

Green, S. (2019, June 7). Let's evaluate Palantir's Berkeley dust-up through Tolkien's lens. The Federalist. https://thefederalist.com/2019/06/07/lets-evaluate-palantirs-berkeley-dust-tolkiens-lens/

Green, V. (2001). *Race on the line: Gender, labor, and technology in the Bell System, 1880–1980*. Duke University Press.

Grinspan, L. (2019, August 24). Many of Miami's immigrants wear ankle monitors. Will technology betray them? *Miami Herald*. https://www.miamiherald.com/news/local/immigration/article234230202.html

Hao, K. (2018, Oct 22). Amazon is the invisible backbone of ICE's immigration crackdown. *Technology Review*. https://www.technologyreview.com/2018/10/22/139639/amazon-is-the-invisible-backbone-behind-ices-immigration-crack-down/

Haskins, C. (2019, July 12). Revealed: This is palantir's top-secret user manual for cops. *Vice*. https://www.vice.com/en/article/9kx4z8/revealed-this-is-palantirs-top-secret-user-manual-for-cops

Hatcher, M. (2017, August 8). 76% of all inmates end up back in jail within 5 years. Here's how I broke the cycle. *Vox*. https://www.vox.com/first-person/2017/8/8/16112864/recidivism-rate-jail-prostitution-break-cycle

Hernandez, C. (2019, October 14). We need more data to understand the impact of mass incarceration on Latinx Communities. Vera Institute of Justice. https://www.vera.org/news/we-need-more-data-to-understand-the-impact-of-mass-incarceration-on-latinx-communities

Hicks, M. (2017). *Programmed inequality: How Britain discarded women technologists and lost its edge in computing*. MIT Press.

Hill, K. (2013, December 19). Data broker was selling lists of rape victims, alcoholics, and "erectile dysfunction sufferers." *Forbes*. https://www.forbes

.com/sites/kashmirhill/2013/12/19/data-broker-was-selling-lists-of-rape-alcoholism-and-erectile-dysfunction-sufferers/?sh=30f9225d1d53

Hill, K. (2020, January 18). The secretive company that might end privacy as we know it. *New York Times*. https://www.nytimes.com/2020/01/18/technology/clearview-privacy-facial-recognition.html

Hrnick. (2017, March 9). The Apple garage. *Atlas Obscura*. https://www.atlasobscura.com/places/apple-garage

Homeland Advanced Recognition Technology. (2021, November 16). *DHS is building a massive database of personal information*. National Immigration Law Center. https://www.nilc.org/wp-content/uploads/2021/12/HART-factsheet-2021–11–10.pdf

Hong, S.-H. (2020). *Technologies of speculation: The limits of knowledge in a data-driven society*. New York University Press.

Igo, S. E. (2007). *The average American: Surveys, citizens, and the making of a mass public*. Harvard University Press.

Igo, S. E. (2018). *The known citizen: A history of privacy in modern America*. Harvard University Press.

Ivanova, I. (2021, March). Immigrant rights groups sue facial recognition company Clearview AI. CBS News. https://www.cbsnews.com/news/clearview-ai-facial-recognition-sued-mijente-norcal-resist

Jaacks, J. (Director). (2020). *Border nation* [Film]. Jason Jaacks Visuals.

Jacobson, M. F. (1998). *Whiteness of a different color: European immigrants and the alchemy of race*. Harvard University Press.

Jain, A. K., Ross, A. A., & Nandakumar, K. (2011). *Introduction to biometrics*. Springer.

Jawetz, T. (2020, September 28). Immigrants as essential workers during COVID-19. *Center for American Progress*. https://www.americanprogress.org/issues/immigration/reports/2020/09/28/490919/immigrants-essential-workers-covid-19/

Jenkings, R. (2012, April 4). How much is your email address worth? *The Drum*. https://www.thedrum.com/opinion/2012/04/04/how-much-your-email-address-worth

Johnson, K. (2021, May 10). Black and queer AI groups say they'll spurn Google funding. *Wired*. https://www.wired.com/story/black-queer-ai-groups-spurn-google-funding/

Juárez, M. (2017, December 26). Maria Teresa Márquez and CHICLE: The first Chicanx electronic mailing list. *Medium*. Retrieved September 15, 2020, from https://migueljuarez.medium.com/mar%C3%ADa-teresa-márquez-and-chicle-the-first-chicanx-electronic-mailing-list-521be58df9db

Junger, S. (2016). *Tribe: On homecoming and belonging*. Hachette Book Group.

Kaplan, A., & Swales, V. (2019, May 6). "Happy hunting": Border patrol searches on buses increase. *Euronews*. https://www.euronews.com/2019/06/05/border-patrol-searches-have-increased-greyhound-other-buses-far-border-n1012596

Kimery, A. (2019, December 16). Army VR terrain tool augmented for border patrol agents. *ARVR Journey*. https://arvrjourney.com/army-reality-terrain-tool-augmented-repurposed-for- border-patrol-agents-d01c3d6100e7

Lambert, K. E. (2018, November). BodyMap creates the "Google map" of the human body. HealthiAR. https://healthiar.com/bodymap-creates-the-google-map-of-the-human-body

Lamdan, S. (2019, November 13). Librarianship at the crossroads of ICE surveillance. *In the Library with the Lead Pipe*. https://www.inthelibrarywiththeleadpipe.org/2019/ice-surveillance/

Lamdan, S. (2022). *Data cartels: The companies that control and monopolize our information*. Stanford University Press.

Lazer, D., Pentland, A., Adamic, A., Aral, S., László-Barabási, A., Brewer, D., Christakis, N., Contractor, N., & Fowler, J. (2009, February). Computational social science. *Science, 323*(5915), 721–723.

Leetaru, K. (2010). The scope of FBIS and BBC open-source media coverage, 1979–2008. *Studies in Intelligence, 54*(1), 51–71.

Leetaru, K. (2011). Culturomics 2.0: Forecasting large-scale human behavior using global news media tone in time and space. *First Monday*. http://firstmonday.org/ojs/index.php/fm/article/view/3663/3040

Lemoult, C. (2022, July 11). When it comes to darker skin, pulse oximeters fall short. NPR. https://www.npr.org/sections/health-shots/2022/07/11/1110370384/when-it-comes-to-darker-skin-pulse-oximeters-fall-short

Leonard, D. (2009). Young, black (or brown), and don't give a fuck: Virtual gangstas in the era of state violence. *Critical Methodologies, 9*(2), 248–272.

Levy, A. (2020, August 25). Palantir CEO rips Silicon Valley in letter to investors. CNBC. https://www.cnbc.com/2020/08/25/palantir-ceo-rips-silicon-valley-in-letter-to-investors.html

Lewis, T., Gangadharan, S. P., Saba, M., & Petty, T. (2018). Digital defense playbook: Community power tools for reclaiming data. *Our Data Bodies*. https://www.odbproject.org/tools/

Library Freedom Project. (n.d.). *Quick courses to sharpen your privacy skills*. https://libraryfreedom.org/crashcourse/

Liljefors, M. & Lee-Morrison, L. (2015). Mapped bodies: Notes on the use of biometrics in geopolitical contexts. In Anders Michelsen and Frederik Tygstrup (Eds.), *Socioaesthetics: Ambience—Imaginary* (pp. 53–72). Brill.

Llamas-Rodriguez, J. (2021). First-person shooters, tunnel warfare, and the racial infrastructures of the US-Mexico Border. *Lateral: Journal of the Cultural Studies Association, 10*(2). https://www.jstor.org/stable/48671648

Luckey, P. [@PalmerLuckey]. (2020, July 3). Immigration policy is a hotly debated issue . . . [Tweet]. Twitter. https://twitter.com/palmerluckey/status/1279300997001564161?lang=en

Lytle Hernández, K. (2010). *Migra! A history of the U.S. Border Patrol*. University of California Press.

Magnet, S. (2006). Playing at colonization: Interpreting imaginary landscapes in the video game tropico. *Journal of Communication Inquiry, 30*(2), 142–162.

Magnet, S. (2011). *When biometrics fail: Gender, race, and the technology of identity*. Duke University Press.

Mai, C. & Subramanian, R.. (2015). Prison spending in 2015. Vera Institute of Justice. https://www.vera.org/publications/price-of-prisons-2015-state-spending-trends/price-of-prisons-2015-state-spending-trends/price-of-prisons-2015-state-spending-trends-prison-spending

Manthrope, R. (2016, September 22). Palmer Luckey's building a VR utopia. And he's only 24. *Wired.* https://www.wired.co.uk/article/palmer-luckey-oculus-rift-virtual-reality-uk-launch

Martínez, O. (1994). *Border people.* University of Arizona Press.

Mayhew, S. (2018, February 1). History of biometrics. Biometric Update website. https://www.biometricupdate.com/201802/history-of-biometrics-2

McAllister, T. (2011, October 11). Update: Bill ends e-verify in Murrieta, other CA cities. *Patch.* https://patch.com/california/murrieta/bill-ends-e-verify-in-murrieta-other-ca-cities

McBride, S. (2013, September 12). Insight: In Silicon Valley start-up world, pedigree counts. Reuters.com. https://www.reuters.com/article/us-usa-startup-connections-insight/insight-in-silicon-valley-start-up-world-pedigree-counts-idINBRE98B15U20130912

McHenry County Historical Society and Museum. (n.d.). RR Boxcar Communities. Retrieved August 5, 2021, from https://mchenrycountyhistory.org/rr-boxcar-communities

McPherson, T. (2012). Why are the digital humanities so white? Or thinking the histories of race and computation. In Matthew K. Gold (Ed.), *Debates in the digital humanities* (part 3, chapter 9). University of Minnesota Press. https://dhdebates.gc.cuny.edu/read/untitled-88c11800–9446–469b-a3be-3fdb36bfbd1e/section/20df8acd-9ab9–4f35–8a5d-e91aa5f4a0ea

Merchant, B. (2017, June 18). Life and death in Apple's forbidden city. *The Guardian.* https://www.theguardian.com/technology/2017/jun/18/foxconn-life-death-forbidden-city-longhua-suicide-apple-iphone-brian-merchant-one-device-extract

Mijente. (n.d.). Blueprint for terror. Retrieved September 20, 2021, from https://mijente.net/icepapers/

Mijente. (2017, January 27). *What makes a sanctuary city now?* https://mijente.net/2017/01/sanctuary-report/

Mijente. (2022, February 1). *Tech wars.* https://instituto.mijente.net/courses/tech-wars/

Molnar, P., & Muscati, S. (2018, September 27). What if an algorithm decided whether you could not stay in Canda or not? *The New Humanitarian.* https://deeply.thenewhumanitarian.org/refugees/community/2018/09/27/what-if-an-algorithm-decided-whether-you-could-stay-in-canada-or-not

Muñiz, A. (2022). *Borderland Circuitry: Immigration Surveillance in the United States and Beyond.* University of California Press.

Nakamura, L. (2014). Indigenous circuits: Navajo women and the racialization of early electronic manufacture. *American Quarterly, 66*(4), 919–941.

National Association for the Advancement of Colored People. Criminal justice fact sheet. https://www.naacp.org/criminal-justice-fact-sheet/

National Immigration Project, Immigrant Defense Project, & Mijente. (2019). Who's behind ICE: The tech and data companies fueling deportations. https://mijente.net/wp-content/uploads/2018/10/WHO'S-BEHIND-ICE_-The-Tech-and-Data-Companies-Fueling-Deportations-_v1.pdf

Naughton, J. (2016, May). The only living Trump supporter in Silicon Valley. *The Guardian*. https://www.theguardian.com/commentisfree/2016/may/22/peter-thiel-paypal-donald-trump-silicon-valley-libertarian

Negroponte, N. (1995). *Being digital*. Hodder and Stoughton.

Ngai, M. (2004). *Impossible subjects*. Princeton University Press.

Noble, S. (2018). *Algorithms of oppression*. NYU Press.

Noble, S. (2021, May 17). *Conversation with Dr. Safiya Noble*. [Conference presentation]. 10th Annual Allen Smith Symposium. https://www.youtube.com/watch?v=oiZJmTtuPlg

Noble, S. and Roberts, S. T. (2019). Technological elites, the meritocracy, and postracial myths in Silicon Valley. In R. Mukherjee, S. Banet-Weiser, & H. Gray (Eds.), *Racism postrace* (pp. 113–134). Duke University Press.

North American Production Sharing, Inc.. Mexico manufacturing specialists. Retrieved December 1, 2022, from https://napsintl.com

O'Flaherty, K. (2019, April). Amazon staff are listening to Alexa conversations—here's what to do. *Forbes*. https://www.forbes.com/sites/kateoflahertyuk/2019/04/12/amazon-staff-are-listening-to-alexa-conversations-heres-what-to-do/?sh=4bc1d79171a2

Ogbar, J. (2017, January 16). The FBI's war on civil rights leaders. *The Daily Beast*. https://www.thedailybeast.com/the-fbis-war-on-civil-rights-leaders

Olson, H. A. (1998). Mapping beyond Dewey's boundaries: Constructing classificatory space for marginalized knowledge domains. *Library Trends, 47*(2), 233–254.

Olson, H. A. (2002). *The power to name: Locating the limits of subject representation in libraries*. Kluwer Academic Publishers.

Olwagen, W.-J. (2015, February 16). From data to information: Introducing information science [Film]. https://www.youtube.com/watch?v=l4sPBKmyWY4

O'Neil, C. (2016). *Weapons of math destruction: How big data increases inequality and threatens democracy* (1st ed.). Crown.

Our Data Bodies Project. (n.d.). What we are doing. https://www.odbproject.org/about/what-we-are-doing/

Oxford Learner's Dictionary. (n.d.). Data mining. In *Oxford Learner's Dictionary*. Retrieved November 21, 2022, from https://www.oxfordlearnersdictionaries.com/us/definition/american_english/data-mining

Palantir. (n.d.). Simple search: Tips and tricks. *Motherboard*. https://www.documentcloud.org/documents/6190005-PALANTIR-Guide.html

Palantíri. (n.d.). *The Lord of the Rings Wiki*. https://lotr.fandom.com/wiki/Palant%C3%ADri

Paredes, A. 1993. "The folklore of groups of Mexican origin in the United States." In Richard Bauman (Ed.), *Folklore and Culture on the Texas-Mexican Border* (pp. 3–18). University of Texas at Austin, Center for Mexican American Studies.

Parrish, W. (2019, August 25). The U.S. border patrol and an Israeli military contractor are putting a Native American reservation under "persistent surveillance." *The Intercept*. https://theintercept.com/2019/08/25/border-patrol-israel-elbit-surveillance/

Patel, F., Levinson-Waldman, R., DenUyl, S. & Koreh, R. (2019). Social media monitoring: How the Department of Homeland Security uses digital data in the name of national security. Brennan Center for Justice at New York University School of Law. https://www.brennancenter.org/our-work/research-reports/social-media-monitoring

Peña, D. G. (1997). *The terror of the machine: Technology, work, gender, and ecology on the U.S.-Mexico border.* University of Texas Press.

Pettitt, J., Malter, J., & Black, E. (2018, December 14). These virtual walls could be the cheaper and more effective answer to Trump's $5 billion border wall. CNBC. https://www.cnbc.com/video/2018/12/14/this-border-town-doesnt-want-trumps-wall-but-a-silicon-valley-virtual-wall-could-stand-strong.html

Phillips, A. (2017, June 16). 'They're rapists.' President Trump's campaign launch speech two years later, annotated. *The Washington Post.* https://www.washingtonpost.com/news/the-fix/wp/2017/06/16/theyre-rapists-presidents-trump-campaign-launch-speech-two-years-later-annotated/

Piven, B. (2019, July 16). What is Amazon's role in the US immigration crackdown? Al Jazeera. https://www.aljazeera.com/economy/2019/7/16/what-is-amazons-role-in-the-us-immigration-crackdown

Preston, J. (2011, January 14). Homeland security cancels "virtual fence" after $1 billion is spent.' *The New York Times.* https://www.nytimes.com/2011/01/15/us/politics/15fence.html

Puar, J. (2007). *Terrorist assemblages: Homonationalism in queer times.* Duke University Press.

Purchese, R. (2013, July 11). Happy go Luckey: Meet the 20-year-old creator of Oculus Rift. *Eurogamer.* https://www.eurogamer.net/articles/2013–07–11-happy-go-luckey-meet-the-20-year-old-creator-of-oculus-rift

Quenya. (n.d.). *The Lord of the Rings Wiki.* https://lotr.fandom.com/wiki/Quenya

Quora contributor. (2012, July 6). What Is the Working Culture Like at Palantir? *Forbes.* https://www.forbes.com/sites/quora/2012/07/06/what-is-the-working-culture-like-at-palantir/#5252c9fb2bafimmigration-crackdown

Rankin, J. (2018). *A people's history of computing in the United States.* Harvard University Press.

Reagon, T., & brown, a. m. (2020, June 22). Parable of the Sower: Chapter 1 [Audio podcast episode]. In *Octavia's Parables.* https://anchor.fm/oparables/episodes/

Reeve, R. G. [@RobertGReeve]. (2021, May 24). [Tweets]. Twitter. Retrieved October 20, 2021, from https://twitter.com/RobertGReeve/status/1397032784703655938

Reilly, K. (2016, August). Here are all the times Donald Trump insulted Mexico. *Time.* https://time.com/4473972/donald-trump-&/

Rivlin-Nadler, M. (2019, December 22). How ICE uses social media to surveil and arrest immigrants. *The Intercept.* https://theintercept.com/2019/12/22/ice-social-media-surveillance/

Roberts, S. (2019). *Behind the screen: Content moderation in the shadows of social media.* Yale University Press.

Rohrlick, J. (2019, November). Homeland security will soon have biometric data on nearly 260 million people. *Quartz*. https://qz.com/1744400/dhs-expected-to-have-biometrics-on-260-million-people-by-2022/

Rollins, P. (2016). *The orthodox heretic and other impossible tales*. Paraclete Press.

Root, N. (2004, June 9). "Accenture faces daunting task with US-VISIT contract." http://wwwinsideid.com.

Saba, M., Lewis, T., Petty, T., Peña Gangadharan, S., & Eubanks, V. (2016). *From paranoia to power*. Our Data Bodies Project. https://www.odbproject.org/wpcontent/uploads/2016/12/ODB-Community-Report-7–24.pdf

Sainato, M. (2019, October 18). "Go back to work": Outcry over deaths on Amazon's warehouse floor. *The Guardian*. https://www.theguardian.com/technology/2019/oct/17/amazon-warehouse-worker-deaths

Sanders, E. & Kilgore, J. (2018, August 4). Ankle monitors aren't humane. They're another kind of jail. *Wired*. https://www.wired.com/story/opinion-ankle-monitors-are-another-kind-of-jail/

Santos, F. (2016, September 12). The time I went on a border patrol in a virtual reality world. *New York Times*. https://www.nytimes.com/2016/09/13/us/the-time-i-went-on-border-patrol-in-a-virtual-reality-world.html

Schuba, T. (2020, January 29). CPD using controversial facial recognition program that scans billions of photos from Facebook, other sites. *Chicago Sun-Times*. https://chicago.suntimes.com/crime/2020/1/29/21080729/clearview-ai-facial-recognition-chicago-police-cpd

Segura, D. A., & Zavella, P. (Eds.). (2007). Introduction. In *Women and migration in the U.S.-Mexico borderlands*. Duke University Press.

Sen, S. (2018). Not quite not white: Losing and finding race in America. Penguin Books.

Shankland, S. (2020, September 10). Oculus founder's ghost 4 military drones use AI for surveillance and attack. CNET. https://www.cnet.com/science/palmer-luckey-ghost-4-military-drones-can-swarm-into-an-ai-surveillance-system/

Sherman, E. (2022, February 1). What Zillow's failed algorithm means for the future of data science. *Fortune*. https://fortune.com/education/business/articles/2022/02/01/what-zillows-failed-algorithm-means-for-the-future-of-data-science/

Singh, R., Guzmán, R.L., and Davidson, P. eds. (2022, Dec 7). *Parables of AI in/from the Majority World*. Data & Society Research Institute. https://datasociety.net/library/parables-of-ai-in-from-the-majority-world-an-anthology/

Silva, D. (2019, August 15). GPS tracking of immigrants in ICE raids troubles advocates. NBC. https://www.nbcnews.com/news/us-news/gps-tracking-immigrants-ice-raids-troubles-advocates-n1042846

Simonite, T. (2014, January 22). Software that augments human thinking: How chess and financial fraud led Palantir to human-machine symbiosis. *Technology Review*. https://www.technologyreview.com/2014/01/22/174482/software-that-augments-human-thinking/

Singh, S. (2019). Love in the time of surveillance capitalism: How algorithms are reshaping our intimate online spaces [Conference session]. Proceedings of 4S Annual Meeting, New Orleans, LA, United States. https://www.4s2019.org/program/

Sohail, U. (2020, February 20). This is a story of life and death in Apple's forbidden city you never heard of. *Wonderful Engineering*. https://wonderfulengineering.com/story-life-death-apple-forbidden-city/

Song, Z. E., & Vázquez, J. (2021, February). Study shows rise of hate crimes, violence against Asian Americans during the pandemic. NBC News. https://www.nbcnewyork.com/news/local/crime-and-courts/study-shows-rise-of-hate-crimes-violence-against-asian-americans-in-nyc-during-covid/2883215/

Southwestern Border Sheriff's Coalition (SBSC) to immediately begin improving. (2017, April). *Business Wire*. https://www.businesswire.com/news/home/20170406006132/en/Southwestern-Border-Sheriffs'-Coalition-SBSC-to-immediately-begin-improving-the-biometric-identification-capabilities-of-the-31-Sheriffs'-Offices-along-the-U.S.-and-Mexico-Border-to-increase-border-security-and-combat-criminal-activity

Spade, D. (2009, February 9). Trans politics on a neoliberal landscape. Barnard Center for Research on Women. https://www.youtube.com/watch?v=oi1fREeZXPI

Spade, D. (2011). *Normal life: Administrative violence, critical trans politics, and the limits of law.* South End Press.

Spade, D. (2020). *Mutual aid: Building solidarity during this crisis.* Verso Press.

Stalder, F. (2002). Privacy is not the antidote to surveillance. *Surveillance & Society, 1*(1), 120–124.

Stanfill, M. (2019). *Exploiting fandom: How the media industry seeks to manipulate fans.* University of Iowa Press.

Stein, S. (2019, December 17). VR in the 2010s: My decade with things on my face. *CNET.* https://www.cnet.com/news/arvr-2009–2019-my-decade-with-things-on-my-face/

Stern, A. M. (2005). *Eugenic nation: Faults and frontiers of better breeding in modern America.* University of California Press.

STTC Research. (2017, November 17). *ARES featured highlights* [Film]. YouTube https://www.youtube.com/watch?v=ouqI1tCYL40&list=PLPGdJVRp7dNjfmrpqj_20meCRakLNgCbq

Swartz, A. (2008). *Guerilla open access manifesto.* Github Gist. https://gist.github.com/usmanity/4522840

Shea Swauger. [@SheaSwauger]. (2019, December 13). [Tweets]. Twitter. Retrieved November 1, 2022, from https://twitter.com/SheaSwauger/status/1205591704973103104?s=20

Sweeney, M. E. (2020). Digital assistants. In Thylstrup, N. B., Agostinho, D., Ring, A., D'Ignazio, C., & Veel, K. (Eds.). *Uncertain archives: Critical keywords for big data.* MIT Press.

Sweeney, M., & Davis, E. (2020). Alexa, are you listening? An exploration of smart voice assistant use and privacy in libraries. *Information Technology and Libraries, 39*(4). https://ejournals.bc.edu/index.php/ital/article/view/12363/10229

Sweeney, M., & Villa-Nicholas, M. (2022). Digitizing the "ideal" Latina information worker. *American Quarterly,* 74(1), 145–167. doi:10.1353/aq.2022.0007.

Tama, J. (2015, July 8). How to get startups in on the military-industrial complex. *Wired*. https://www.wired.com/2015/07/get-startups-military-industrial-complex/

US Customs and Border Protection. (2005, September 9). Border safety initiative. Homeland Security Digital Library. https://www.hsdl.org/?view&did=457099

US Customs and Border Protection. (2019, December 20). Did you know . . . B.Y.O.H.? Early border patrol agents met with stringent requirements. https://www.cbp.gov/about/history/did-you-know/byoh

US Customs and Border Protection. (2020, January). Border patrol sectors. https://www.cbp.gov/border-security/along-us-borders/border-patrol-sectors/san-diego-sector-california/murrieta-station

U.S. Customs and Border Protection. (2020, January 22). Theodore L. Newton, Jr. and George F. Azrak Station. https://www.cbp.gov/border-security/along-us-borders/border-patrol-sectors/san-diego-sector-california/murrieta-station

US Customs and Border Protection. (2022). 1924: Border patrol established. California Border Patrol. Retrieved November 21, 2022, from https://www.cbp.gov/about/history/1924-border-patrol-established

US Customs and Border Protection. (n.d.) Border patrol history. US Customs and border protection. Retrieved November 21, 2022, from https://www.cbp.gov/border-security/along-us-borders/history

US Department of State. (2008, December 4). Fraud in the refugee family reunification (Priority Three) program. https://2001–2009.state.gov/g/prm/refadm/rls/fs/2008/112760.htm

US Government Accountability Office. (2017, November). *Southwest border security: Border patrol is deploying surveillance technologies but needs to improve data quality and assess effectiveness*. GAO-18–119.

US House of Representatives. (1999, November 10). *Recognizing the U.S. border patrol's seventy-five years of service; Congressional record*, 145, no. 158. 106th Congress. https://www.congress.gov/congressional-record/1999/11/10/house-section/article/H11922–1

US Immigration and Customs Enforcement. (n.d.). Fugitive operations. Retrieved June 20, 2020, from https://www.ice.gov/identify-and-arrest/fugitive-operations

USA Today Network. (2018, January 24). The wall VR experience. AZ Central. https://www.azcentral.com/videos/news/politics/border-issues/2019/03/30/-wall-vr-experience/109563720/

Velez, L. (2022). "It was like he was writing my life": How ethnic identity affected one family's interpretation of an Afro Latinx text. *Library Trends, 70*(2), 207–238.

Villa-Nicholas, M. (2015). Latina/o librarian technological engagements: REFORMA in the digital age. *Latino Studies, 13*(4), 542–560. https://doi.org/10.1057/lst.2015.43

Villa-Nicholas, M. (2018, May). Terror by Telephone: Normative anxieties around obscene calls in the 1960s. *First Monday,* 23(5). https://doi.org/10.5210/fm.v22i5.7010

Villa-Nicholas, M. (2022). *Latinas on the line: Invisible information workers in telecommunications*. Rutgers Press.

Villa-Nicholas, M., & Sweeney, M. (2020). Designing the "good citizen" through Latina identity in USCIS's virtual assistant "Emma." *Feminist Media Studies, 20*(7), 909–925.

Volpp, L. (2002). The citizen and the terrorist. *UCLA Law Review, 49*(5), 1575–1600.

Waldman, P., Chapman, L., & Robertson, J. (2018, April 19). Palantir knows everything about you. *Bloomberg*. https://www.bloomberg.com/features/2018-palantir-peter-thiel/

Ward, J., & Sottile, C. (2019, October 3). Inside Anduril, the startup that is building AI-powered military technology. *NBC News*. https://www.nbcnews.com/tech/security/inside-anduril-startup-building-ai-powered-military-technology-n1061771

Watts, J., & Peretti, J. (2007, May). Viral marketing for the real world. *Harvard Business Review*. https://hbr.org/2007/05/viral-marketing-for-the-real-world

Weigel, M. (2020, July 10). Palantir goes to the Frankfurt school. *Boundary2*. https://www.boundary2.org/2020/07/moira-weigel-palantir-goes-to-the-frankfurt-school/

Weiser-Alexander, K. (2020, January). Westward expansion and manifest destiny—Legends of America. *Legends of America*. https://www.legendsofamerica.com/westward-expansion/

Wernimont, J. (2018). *Numbered lives : Life and death in quantum media*. MIT Press.

Wevers, R. (2018). Unmasking biometrics' biases: Facing gender, race, class and ability in biometric data collection. *TMG Journal for Media History, 21*(2), 89–105.

Wilkerson, I. (2020). *Caste: The origins of our discontents* (1st ed.). Random House.

Winner, L. (1986). Do artificants have politics? In L. Winner (Ed.), *The whale and the reactor: A search for limits in an age of high technology* (pp. 19–39). University of Chicago Press.

Winter, J. (2021, May 18). Facial recognition, fake identities, and digital surveillance tools: Inside the post office's covert internet operations program. *Yahoo News*. https://news.yahoo.com/facial-recognition-fake-identities-and-digital-surveillance-tools-inside-the-post-offices-covert-internet-operations-program-214234762.html

Wise, S. & Petras, G. (2018, June). The process of deportation. *USA Today*. https://www.usatoday.com/pages/interactives/graphics/deportation-explainer/

Worth, K. (2015, December). For some refugees, safe haven now depends on a DNA test. PBS. https://www.pbs.org/wgbh/frontline/article/can-biometrics-solve-the-refugee-debate/

Wu, E. D. (2013). *The color of success*. Princeton University Press.

Wyatt, S. (2003). Non-users also matter: The construction of users and non-users of the Internet. In Oudshoorn, N. & Pinch, T. (Eds.), *How users matter: The co-construction of users and technology* (pp. 67–79). MIT Press.

Zak, P. (2011, November). *Trust, morality—and oxytocin?* https://www.ted.com/speakers/paul_zak

Zak, P.J. (2012). *The moral molecule: The source of love and prosperity.* Bantam Books.

Index

Page numbers in italics refer to tables.